Logic, Epistemology, and the Unity of Science

Founding Editor

John Symons

Volume 52

Logic, Epistemology, and the Unity of Science aims to reconsider the question of the unity of science in light of recent developments in logic. At present, no single logical, semantical or methodological framework dominates the philosophy of science. However, the editors of this series believe that formal frameworks, for example, constructive type theory, deontic logics, dialogical logics, epistemic logics, modal logics, and proof-theoretical semantics, have the potential to cast new light on basic issues in the discussion of the unity of science.

This series provides a venue where philosophers and logicians can apply specific systematic and historic insights to fundamental philosophical problems. While the series is open to a wide variety of perspectives, including the study and analysis of argumentation and the critical discussion of the relationship between logic and philosophy of science, the aim is to provide an integrated picture of the scientific enterprise in all its diversity.

This book series is indexed in SCOPUS.

For inquiries and submissions of proposals, authors can contact Christi Lue at christi.lue@springer.com

More information about this series at http://www.springer.com/series/6936

Claudia Fernández-Fernández

Awareness in Logic and Epistemology

A Conceptual Schema and Logical Study of The Underlying Main Epistemic Concepts

Springer

Claudia Fernández-Fernández
Departamento de Filosofía
Universidad de Málaga
Málaga, Spain

ISSN 2214-9775 ISSN 2214-9783 (electronic)
Logic, Epistemology, and the Unity of Science
ISBN 978-3-030-69605-4 ISBN 978-3-030-69606-1 (eBook)
https://doi.org/10.1007/978-3-030-69606-1

This Springer imprint is published by the registered company Springer Nature Switzerland AG
The registered company address is: Gewerbestrasse 11, 6330 Cham, Switzerland

Contrariwise, continued Tweedledee,

if it was so, it might be; and if it were so,

it would be; but as it isn't, it ain't.

That's logic.

Carroll, Lewis (1920). Alice's Adventures in Wonderland.

New York: Macmillan.

Preface

I take the liberty of introducing the main goals of this book with a *metaphoric tale* that will guide you through the different paths that will be covered.

Once upon a time, about thirteen years ago, in a highland, a woman named *Sofia* gave birth to two beautiful twin girls. Sofia was very fond of Aristotle and Greek Philosophy in general, so she named her girls *Episteme* and *Logos*.

Some weeks later, there was a terrible fire that burned down the house and took many lives from the small town in the highland. The only survivors were the twin girls, whose destiny was altered tragically. The State tried to find a new family for the girls, but they were forced to be separated, since the only adoptive families they found could only take one of the girls, respectively. Thus, Episteme stayed near the highland with a humble family, while Logos was adopted by a family who lived in the big city.

Episteme grew up in harmony with nature. Her family were artists and writers, who inspired her an interest for reading, writing and learning foreign languages. By the age of sixteen she was already fluent in Spanish and French and had learned some German and Greek. When she turned eighteen and finished high school she decided to move to the big city and start a Philosophy degree at the university.

Logos, on the other hand, grew up in a technological environment. Her adoptive family were engineers and technicians, who taught her, at a very young age, some programming skills. Thereupon, Logos was attracted to logical puzzles, electrical circuits and computer programming. During her teenage years she enjoyed developing software for making her daily life easier. After finishing high school she decided to study Computer Science at the university.

The degrees of Philosophy and Computer Science share a common subject at this university in the first semester of the first year, namely, *Philosophical Logic*. The first day of class two girls were rushing through the corridor to get to room 314, the Philosophical Logic class. Right at the entrance, before even entering the room, they looked up at each other and their faces went pale. It was as if they were looking at a mirror, but they were not. Their faces looked exactly the same, though their hairs and make-up were different.

'Who are you?', said Episteme after starring at Logos for a long time. 'And who are *you*?', replied Logos. 'Well, I am Episteme from the highland', Episteme said. 'And I am Logos from the city. But how is this possible?', answered Logos looking puzzled. From that moment on, the twin girls began getting to know each other. They obviously did not know that they had a twin sister and, as it turned out, neither did their adoptive families.

Episteme and Logos spent most of the following days together, learning about each other's life. They soon noticed that though they had different interests, their characters had many similarities and they were both intellectually curious. But, when Logos started telling Episteme what her hobbies were, the communication began to fail since they did not share a common background knowledge.

Episteme was unable to understand, at first, all those strange programming languages and concepts that Logos tried to explain to her. And also the other way round, when Episteme told Logos about her interest in Philosophy and mentioned some authors and theories, Logos was astonished. She had no clue what her sister was talking about, though she was speaking in English.

This initial state of general confusion changed very quickly, once both caught up with each other's hobbies and its specific vocabularies. At this point, they were able not only to understand everything the other sister said, but also to find some common ground, thanks also to the subject of Philosophical Logic, which they enjoyed very much.

As they both grew older, finished their degrees and started their respective academic careers, they decided to collaborate and deepen into the connections between Epistemology and Epistemic Logic, which not only was a *hot topic* at the time, but also served them for giving a sense to their own life experiences.

This tale entails as its moral that though very different in its form, language and methods, the fields of Epistemology and Epistemic Logic share the same root in its foundations and can be reconnected at some point. This is precisely the main motivation for this book. When I realised that these fields of study, which I was aware that shared a common ground, had been growing and developing in almost complete isolation from one another, I felt the need to explore its roots and develop a 'tool' for reconnecting them. With the creation of the conceptual schema, that this book provides, the two subjects will find that common ground, though only between two very specific areas (Awareness Justification Internalism and a dynamic interpretation of Awareness Logic).

The core concept through which this bridge will be built and that connects the specific epistemological view with a concrete proposal of Epistemic Logic is that of *Awareness* (in its epistemic interpretations), as will be properly defined and justified. As well in Epistemology as in Epistemic Logic this concept will play a central role. The study of how we think or of how we come to think what we know has to take into account, in one sense or another, everything we consider or entertain. In other words, (i) Epistemology has to deal with Awareness.

In addition to that, the formal representation of our knowledge and informational attitudes would fall short in its connection to human knowledge should it not include Awareness in its system. And that means that (ii) Epistemic Logic has to deal with Awareness.

Claims (i) and (ii) should not be taken as dogmas, but rather as hopes or wishes that are fulfilled with the present research. I do not state that 'awareness' should be part of *every* epistemological view or system of Epistemic Logic. Instead, I claim that under the focus of this research, namely, that of reconnecting Epistemology and Epistemic Logic based on an intuitive understanding of the informational processes in human beings, the notion of 'Awareness' (or, at least, a similar concept) needs to be taken into account.

By analysing, classifying and combining the different interpretations of 'Awareness' I will build a theoretical bridge that serves Epistemic Logic as its philosophical foundation and Epistemology as its logical correlate. The bridge will constitute a conceptual schema that unifies, relates and connects the main topics of both fields. It also serves to enrich these fields, creating thus a virtuous circle. Finally, I will present a formal model that puts all these concepts together and formalises not only the core notions, but also their dynamics, representing epistemic actions that change the available information.

This book is the result of a revision of my Ph.D. Thesis, which I defended on 30th September 2019 in Málaga (Spain). My acknowledgement goes to every single being that has ever crossed my path and left a mark or memory. I like to live every life experience as an opportunity to learn and make it all worthwhile. Regarding the creation of this book, I want to thank Alfredo Burrieza, my supervisor and advisor; Fernando Velázquez-Quesada, my co-author, unofficial advisor and colleague; the members of the Doctoral Thesis Tribunal and reviewers, who encouraged me to prepare this book; Pedro Chamizo, whose valuable advices and reviews are always enriching; and my family, for their love and support.

Málaga, Spain
November 2020

Claudia Fernández- Fernández

About This Book

This book explores the current state of affairs of the relation between mainstream Epistemology and Epistemic Logic. The apparent disconnection between these fields of study is reverted with the creation of a conceptual schema that acts as a theoretical bridge between both areas. With the concept of (epistemic) awareness as a guideline, it offers a new understanding of the main epistemic concepts from Epistemology and Epistemic Logic, where the notion of Explicit Aware Knowledge (EAK) plays a central role.

EAK is presented as the (propositional) knowledge that Epistemology is interested in, as well as the type of knowledge than an extension of Epistemic Logic should represent. The relevance of the concept of 'awareness' is shown with respect to three different applications: contemporary views on Epistemic Internalism, the dynamics of information in logical proposals and a formal model that represents the different informational attitudes of the agent, including the epistemic actions that formalise the changes in information.

The interdisciplinary nature of this book makes it neither a book only on Epistemology, nor only about Epistemic Logic, nor about the dynamics of information, though it is a book about all of these topics. The content can be classified between Philosophy of Logic and mainstream Epistemology. This book discusses the epistemic concepts of *awareness*, *knowledge* and *justification*, and proposes some new theoretical connections and re-definitions.

The structure of this book is the following one: Chap. 1 is devoted to an introduction to the main topics in Epistemic Logic and Epistemology and reviews the disconnection between these fields of study. Chapter 2 reviews the concept of 'Epistemic Awareness', distinguishing between 'Awareness-of' and 'Awareness-that'. Awareness-of represents the attention of the agent, analogous to the language she has at her disposal at a given moment and upon which she can create new information. Awareness-that stands for the acknowledgement of the truth of some information, being thus, a type of knowledge. Chapter 3 considers the notion of 'Knowledge' and analyses its different interpretations in Epistemology and some contemporary developments of Epistemic Logic. It draws a distinction between implicit knowledge, representing an 'ideal' understanding of knowledge, which is the result of the closure under logical consequence of explicit knowledge; and the 'real' knowledge, which Epistemology is about and that represents the human knowledge, which is named 'Explicit Aware Knowledge' (EAK). The EAK is formed due to a combination of both types of awareness, such that for an agent to explicitly know some given information she needs to be aware-that it is the case and aware-of this information at a given moment. Hereafter, Chap. 4 reviews the different interpretations of the concept of 'Justification' and it is argued that justification will be defined as the process through which the agent obtains her EAK. This process is given by the different epistemic actions that are included in the conceptual schema and two of them will be able to provide a justification. Due to an act of deductive inference the agent will obtain a deductive or inferential justification, while an act of observation (or communication) will provide her with an evidential or observational justification.

Finally, Chap. 5 presents the EAK-Schema with the visual help of a diagram formed by three ellipses. The diagram starts out with a central small ellipse representing the Awareness-that, those pieces of information the agent has acknowledged as true at any moment. From there, the Implicit Awareness-that can be deductively inferred, representing thus a bigger ellipse, containing the first one. As a third step, it incorporates the concept of Awareness-of as another ellipse, of the same size as the implicit information, overlapping the previous two ellipses, but not completely, creating thus three parts of Awareness-of. All these parts represent information the agent is considering at a given moment: pure Awareness-of (where there is space for any type of information, even false one), Implicit Knowledge and Explicit Aware Knowledge, EAK (as the only 'real' knowledge). EAK is located at the centre of the diagram, representing the kernel of it and corresponding also to the only type of knowledge that Epistemology will consider. With the visual help of the diagram and some arrows, epistemic actions that transform the information available for the agent (following the tradition of Dynamic Epistemic Logic) are presented. Acts of becoming aware-of or becoming unaware-of will change the information of the Awareness-of, modifying what the agent is paying attention to. The actions of performing a deductive inference or forgetting will either increase or decrease the information the agent has acknowledged, that is, her Awareness-that. Lastly, the action of observation (or communication in multi-agent settings) allows any piece of information to be directly part of the EAK. On the one hand, the action of deductive inference (transforming implicit information into EAK) constitutes the process for obtaining a deductive justification. An act of observation, on the other hand, will provide the agent's EAK with an observational justification, indicating the source of its knowledge.

Chapter 6 presents a formal model, based on a neighbourhood-model structure, that shows one concrete application of the EAK-Schema into a logical structure. The Awareness-of is represented syntactically with a global awareness-function, while Awareness-that is depicted semantically due to the neighbourhood function. The concepts of Explicit Aware Knowledge and Implicit Knowledge are defined, and a list of their properties is provided. In addition, it includes a formalisation of the five epistemic actions defined in the EAK-schema and analyses some of its main features regarding the Explicit Aware Knowledge. To end this book, Chap. 7 is devoted to the conclusions and final remarks about the EAK-Schema, its uses and applications. The most important feature is that it serves both as a theoretical correlate to the dynamic extensions of Awareness Logic, providing it with a philosophical background, and as an abstract conceptual structure for a re-interpretation of Epistemology (specifically, for the Awareness Justification Internalism).

Contents

About the Author

Claudia Fernández-Fernández is graduated in Philosophy (2011) by the University of Málaga (Spain), Master in Analytic Philosophy (2013) by the University of Barcelona (Spain) and Ph.D. in Philosophy (2019) by the University of Málaga (Spain), with a dissertation on Knowledge and Awareness in Logic and Epistemology. Her research interests are related to Logic, Philosophy of Logic, Philosophy of Mathematics and Philosophy of Language.

Since 2015 she works as a substitute interim teacher at the Department of Philosophy in Málaga University (Spain), teaching courses related to Philosophy of Logic, Philosophy of Mathematics and Semantics. Most of her publications are related to the topic of her Ph.D. dissertation, such as Fernández-Fernández, C. and Velázquez-Quesada, F. R. (2020). Awareness-of and awareness-that their combination and dynamics. *Logic Journal of the IGPL*, DOI: https://doi.org/10.1093/jigpal/jzz043; Fernández-Fernández, C. (2019). Awareness Logic: an Epistemological Defence. Correlations between Awareness Logic and Epistemology, *Kairos. Journal of Philosophy &Science*, 22(1), 72–85. DOI: https://doi.org/10.2478/kjps-2019-0009; and Burrieza, A. and Fernández-Fernández, C. (2015). Distributed Explicit Knowledge and Collective Awareness. *Bulletin of the European Association for Theoretical Computer Science*, Number 116 (June 2015), pp. 195–203.

She has presented her work on several international conferences such as *XX Model-Based Reasoning in Science and Technology. Inferential Models for Logic, Language, Cognition and Computation*, October 2018 in Seville (Spain); *The 18th Trends in Logic International Conference*, September 2018 in Milan (Italy); *Logica 2018*, June 2018 in Hejnice (Czech Republic); *Logica 2017*, June 2017 in Hejnice (Czech Republic) and *Third Lisbon International Conference on Philosophy of Science: contemporary issues*, December 2016 in Lisbon (Portugal).

Abbreviations

AGM	Alchourrón, Gärdenfors and Makinson
AI	Artificial Intelligence
AJI	Awareness Justification Internalism
AL	Awareness Logic
ANM	Awareness Neighbourhood Model
DEL	Dynamic Epistemic Logic
EAK	Explicit Aware Knowledge
EL	Epistemic Logic
JL	Justification Logic
JTB	Justified True Belief
LP	Logic of Proof
MWD	Merriam-Webster Dictionary
OD	Oxford Dictionary
PAL	Public Announcement Logic

List of Figures

Chapter 1
Introduction to Epistemic Logic and Epistemology

Abstract This chapter summarizes the most relevant preliminary concepts from Epistemic Logic and Epistemology. Regarding the former, the problem of logical omniscience and its different solutions are revised, with Awareness Logic as the chosen solution for carrying out this research. With respect to Epistemology, a brief historical background is presented and the view of Awareness Justification Internalism is highlighted as the one that will serve as a theoretical background for the logical approach. To conclude, some aspects of the disconnection between both disciplines are pointed out and the concept of Epistemic Awareness is defined as the bridge-notion through wich the theoretical re-connection will be developed.

Keywords Epistemic logic · Epistemology · Epistemic internalism · Awareness logic · Logical omniscience

1.1 Epistemic Logic: Main Concepts

This section provides the reader with a brief and sometimes schematic presentation of the main logical concepts that will be employed during the following chapters. While introducing these main notions some of them will be highlighted as significantly relevant; and in those cases where it comes down to different interpretations some choices will be made and justified.

Epistemic Logic. Epistemic Logic (EL) was 'born' in von Wright (1951). Von Wright took the first steps towards a formal study of knowledge and belief. Some years later, in Hintikka (1962), the author transforms von Wright's ideas into more semantic concepts and EL was established as what we know today. EL has been developed and revisited by many authors since then.

Epistemic Logic[1] is an extension of Modal Logic, in which the necessity operator ($\Box$) is interpreted as knowledge and ranges over a set of epistemic agents ($\mathtt{Ag}$),[2]

[1]What follows is one possible contemporary presentation of EL.

[2]In *doxastic logic*, the logic of belief, the necessity operator is interpreted as what the agent believes.

C. Fernández-Fernández, *Awareness in Logic and Epistemology*,
Logic, Epistemology, and the Unity of Science 52,
https://doi.org/10.1007/978-3-030-69606-1_1

such that the formula $\Box_i\varphi$ is read as *agent* i *knows* φ (or 'agent i knows the truth of φ' or 'agent i knows that φ is the case'). The semantic structure of EL, like Modal Logic, is based on the *possible worlds model* (also called Kripke model), where every epistemic possibility (possible world) an agent may consider is represented via the epistemic accessibility relation of each agent to those possible worlds she considers.

An **epistemic model** is the structure in which the knowledge of the agents is represented. Let $\mathtt{P}$ be a set of atomic propositions and let $\mathtt{Ag}$ be a set of epistemic agents. Then an epistemic model is a tuple $\langle W, \{R_i\}_{i\in\mathtt{Ag}}, V\rangle$, such that:

- $W \neq \varnothing$ is a non-empty set of possible worlds,
- $R_i \subseteq (W \times W)$ is the epistemic accessibility relation of agent i (indicating the worlds each agent considers possible from each one of them), and
- $V : W \to \wp(\mathtt{P})$ is the atomic valuation function (indicating the atomic propositions in $\mathtt{P}$ that are true at each possible world).

The pair (M, w), with $w \in W$, is called an **epistemic state in model M** and corresponds to the evaluation point, that is, the world in which the given formula for the given agent will be assessed based on the other worlds that are accessible from w for agent i. Concretely, from each world $w \in W$, the agent considers possible all those worlds u that she can R-access from w.

The epistemic model M is described by the **language** of epistemic logic, $\mathcal{L}(\Box)$, whose formulas φ, ψ are given by:

$$\varphi ::= p \mid \neg\varphi \mid \varphi \wedge \psi \mid \Box_i\varphi, \quad \text{where} \quad p \in \mathtt{P} \text{ and } i \in \mathtt{Ag}.$$

Recall that $\Box_i\varphi$ is read as *agent* i *knows that* φ *is the case*. The other Boolean connectives can be defined from negation ($\neg$) and conjunction ($\wedge$) in the standard way.

The **semantic interpretation** of the formulas is such that for (M, w) being an epistemic state in $M = \langle W, \{R_i\}_{i\in\mathtt{Ag}}, V\rangle$ and φ and ψ any formulas in $\mathcal{L}(\Box)$; $(M, w) \Vdash \varphi$ indicates that φ is true at w in M (and $(M, w) \nVdash \varphi$ indicates that it is not true). Then,

- $(M, w) \Vdash p$ iff $w \in V(p)$
- $(M, w) \Vdash \neg\varphi$ iff $(M, w) \nVdash \varphi$
- $(M, w) \Vdash \varphi \wedge \psi$ iff $(M, w) \Vdash \varphi$ and $(M, w) \Vdash \psi$
- $(M, w) \Vdash \Box_i\varphi$ iff for every $u \in W$, $R_i wu$ implies $(M, u) \Vdash \varphi$.

Note how the semantic interpretation of $\Box_i\varphi$ says that agent i knows φ at w if and only if φ is true at all worlds u that are accessible for i from w due to R_i.

The concepts of **satisfiability** and **validity** are defined in a standard way. Let φ be a formula in $\mathcal{L}(\Box)$. A formula is said to be *satisfiable* if and only if it is true in at least one world w of at least one model M (that is, if and only if there is one M such that $(M, w) \Vdash \varphi$). A formula is said to be *valid* if and only if it is true at all worlds w of all models M. The valid formulas are represented as follows: $\vdash \varphi$.

The **knowledge operator** ($\Box_i$) stems from the universal necessity modal operator. As such, it has the same properties as the modal operator. This means that the knowledge operator satisfies both the *rule of necessity (Nec)*, $\vdash \varphi$ implies $\vdash \Box_i\varphi$, and the

K-axiom, $\vdash \Box_i(\varphi \rightarrow \psi) \rightarrow (\Box_i\varphi \rightarrow \Box_i\psi)$. The first one states that if a formula is true, then agent i knows it. The second property makes the knowledge closed under logical consequence, meaning that if agent i knows φ and $\varphi \rightarrow \psi$ at world w, then she automatically knows ψ via the *Modus Ponens* rule. And this, conversely, means that now ψ is also true at every u accessible from w. This feature of Epistemic Logic has been given the name of the *problem of logical omniscience.*

The problem of logical omniscience. The fact that the knowledge of the epistemic agents that EL proposes is closed under logical consequence should not be a problem by itself. The controversy arises when the notion of 'epistemic agents' is applied to human beings, instead of abstract entities that posses knowledge (like some ideal agents from Artificial Intelligence (AI)). But, of course, when one tries to understand the concept of 'epistemic agent', the most common and intuitive interpretation is to relate it to ourselves.

This was also the case of Hintikka, who in his seminal work argued in favour of Epistemic Logic as a way of representing human knowledge. He was already aware of the difficulties that this would generate and stated that: "[t]he logical implications of what we know do not come to us without any work on our own part; they are truths which we can extract, often with considerable labor, from whatever information we already have" (Hintikka 1962, p. 37). He continued advocating that although the laws of logic are not equivalent to the laws of thought, they could be the "laws of the sharpest possible thought" (Hintikka 1962, p. 37). A few years later, in (Hintikka 1975) he proposed one strategy to solve the logical omniscience of epistemic agents as will be explained below.

Human subjects, though rational, are not omniscient (we could leave this adjective for the knowledge of some God(s)). In fact, humans, as computing machines, have *limited reasoning abilities*. Though the rationality of human subjects has not been questioned, what is clear is that it is limited, as our lives are.

In *Reasoning about Knowledge* Fagin et al. (1995), the authors devote a chapter to the problem of logical omniscience (LO) and its possible solutions.[3] They state that the problem as defined above corresponds to *Full Logical Omniscience*. There are also weaker forms of omniscience derived from the Full LO, such as the closure under logical implication or closure under logical equivalence. Other types of LO, not derivable from the Full LO are also mentioned, like closure under conjunction or closure under valid implication. As may already be clear, all forms of LO are related to some kind of closure property that is imposed on the knowledge of the agents in Epistemic Logic.

As the authors wisely conclude, the different strategies out of LO pass through a reconsideration of the conditions for knowledge or the definition of 'knowledge' itself. In the mentioned chapter (Fagin et al. 1995, pp. 309–362) they make a thorough analysis of the existing solutions to the problem of LO by that time, classifying them with respect to the semantic or syntactic structure that is imposed and the different concepts that are being redefined. A complete overview of these solutions may be

[3]For an updated overview and contemporary proposal about this topic see Hawke et al. (2019).

found in the given reference and would lie beyond the limits of this chapter. Some key notions for a general glimpse should thus suffice.

The **solutions to the problem of LO** need to reconsider the conditions for knowledge to be true. In standard EL knowledge is true at a given world if it is true in all worlds accessible from the given one, and that was exactly the cause of the problem. It is an ideal representation of knowledge. One way of solving the problem would be to establish a *different definition of knowledge*, from a syntactic or a semantic point of view. Though attractive as a solution to the problem, these strategies may also leave knowledge without other properties that were important; they are useful for defining knowledge, but not for modelling it formally.

There are also many *non-standard logics* that redefine knowledge or the whole structure of Epistemic Logic. These non-standard approaches commonly change the notion of truth by extending some operator or some part of the semantic structure with a counterpart where not every property or validity holds. Though interesting and applicable to concrete problems in AI and other fields, the authors do not highlight any of them as a complete solution to the problem of LO (if such 'complete solution' even exists).

The two most promising strategies they present, from an intuitive and theoretical point of view, are *impossible worlds* and *awareness logic*. The impossible worlds approach was proposed in Hintikka (1975). He assumes the problem of LO and proposes to change the semantic structure, inspired by contemporary ideas from Veikko Rantala on surface models in Rantala (1975). Hintikka establishes a difference between an *epistemic alternative* and a *logical possibility*, where the former are all the alternatives the agent may consider (including impossible worlds in which, for example, contradictory information may hold), and the latter are maximally consistent descriptions of the world. Using a complex structure he manages to adapt the semantics of possible worlds for including also impossible worlds, having now the possibility of evaluating formulas in both types of worlds. It can be considered a good solution for the problem of LO, since agents are not logically omniscient with respect to the impossible worlds, but, on the other hand, they can 'know' everything in those worlds, making their knowledge in impossible worlds quite trivial.

I left **Awareness Logic** (AL) until last, since it is the strategy I consider most suitable for the purpose of solving the problem of LO in EL. The main idea behind this approach, already proposed in Fagin and Halpern (1988), is that in order to know something, it is necessary to be aware of it. They state it in the following terms:

> The underlying idea is that it is necessary to be *aware* of a concept before one can have beliefs about it. One cannot know something of which one is unaware. Indeed, how can someone say that he knows or doesn't know about p if p is a concept of which he is completely unaware? (Fagin et al. 1995, p. 337. Original italics).

With this view in mind, they propose to add awareness to the standard system of EL as a syntactic operator, acting as a filter on the knowledge of the agent. They also incorporate a new operator for the *explicit knowledge*, representing the 'real' knowledge the agent possesses. The explicit knowledge of the agent is then defined by the conjunction of both implicit knowledge (standard knowledge in EL) and awareness.

Awareness is defined as an arbitrary list of formulas and only those that coincide with the ones implicitly known, will be the explicitly known formulas. Formally, they modify the original language $\mathcal{L}(\Box)$, being now $\mathcal{L}^A(\Box)$, formed by $\mathcal{L}(\Box)$ and two new operators: the *awareness operator*, A, such that formulas of the form $\mathrm{A}_i\,\varphi$ are read as *agent i is aware of* φ; and the *explicit knowledge operator*, X, with formulas of the form $X_i\varphi$ read as *agent i knows* φ *explicitly*.

The semantic model M is modified, adding the *awareness function*, $\mathfrak{A}$, that assigns a list of formulas, of which each agent is aware of, to each world. Then M^A is an *awareness model*: $M^A = \langle W, \{R_i\}_{i\in \mathsf{Ag}}, V, \mathfrak{A}_i \rangle$.

The *semantic interpretations* for the two new operators are the following ones: let (M^A, w) be an awareness epistemic state with $w \in W$ and φ any formula in $\mathcal{L}^A(\Box)$, then

- $(M^A, w) \Vdash \mathrm{A}_i\,\varphi$ iff $\varphi \in \mathfrak{A}_i(w)$
- $(M^A, w) \Vdash X_i\varphi$ iff $(M^A, w) \Vdash \mathrm{A}_i\,\varphi$ and $(M^A, w) \Vdash \Box_i\varphi$

From the latter, the authors conclude the following validity, which is the key of this proposal: $-\ X_i\varphi \leftrightarrow (\mathrm{A}_i\,\varphi \wedge \Box_i\varphi)$, stating the logical equivalence between agent i having explicit knowledge about φ and her fulfilling both implicitly knowing φ *and* being aware of φ.

The original proposal leaves the awareness function without any restrictions, but they argue that different closure properties might easily be achieved, such as closure under subformulas or self-reflection. This proposal does overcome the problem of LO and maintains the original structure of EL. It is also compatible with the previous proposal of impossible worlds and with some other semantic structures. Probably the stroke of genius that comes with this proposal, is that Fagin and Halpern managed to formalize a common sense intuition that was already present in many other fields of study.

The potential that AL has, was fruitfully explored in the following decades and incorporated to the new paradigm of the *studies of language and information*. There are also other proposals that include an external factor to the main structure of EL in order to divide knowledge into implicit and explicit and hence prevent the agents from being logically omniscient, but I will not review them here for lack of interest to this research. Though the authors present it as one possible solution among others, they had already explored this approach (as mentioned before) and continued afterwards this line of research.

New approaches of knowledge and information. The last decades have been very prolific in the studies of language and information from the logical point of view. One could speak of a new *paradigm* that has been settled regarding the logical interactions of language and information (and also computation).

The different ways in which information is represented will give rise to the different logics and semantics that one finds nowadays. The milestone that turned the concepts from Epistemic Logic into a whole new paradigm was probably the concept of *information change*. In contemporary approaches it is hard to find the mere concepts of 'knowledge' and 'belief' as such. Instead, one finds the notion of 'information' and its various interpretations will turn it into either knowledge, or belief, or other types of epistemic attitudes. Of course the concept of 'information change' has its origins in the studies of communication, that are seen as multi-agent scenarios in Epistemic Logic, where the given formulas or propositions change during the communication process.

Epistemic Logic represents a simple semantic interpretation of information. The new approaches to the notion of information (change) have been made from a syntactic and semantic point of view. The purely *syntactic representations* understand 'information' as a list of formulas, or propositions, and sometimes add some properties to the set of information that is assigned to an agent for achieving (or coming closer to) omniscience. The interactions between these formulas or propositions are then deductive. There are different proposals, with *Belief Revision* being one of the most relevant ones (for a general overview see Gärdenfors (1992) and Williams and Rott (2001), for example).

In **Belief Revision Theory** one could highlight the *AGM-model* (named after the authors Alchourrón, Gärdenfors and Makinson that published the originating paper Alchourrón et al. 1985). This proposal of representation of belief establishes a distinction between the *belief set* (consistent set of formulas closed, normally, under logical consequence) and the *belief base* (simple set of formulas acting as a basis for the belief set, without closure properties). The AGM-model provides tools for formalizing the different types of changes (e.g., revisions and updates) there might be in the agent's belief set, but from a purely syntactic point of view. As pointed out in (Velázquez-Quesada 2011, p. 9), "syntactic approaches have been criticized as being too fine-grained, making differences in meaning where there seems to be none". However, Belief Revision Theory had great influence in the forthcoming approaches and is still very useful in different fields of Artificial Intelligence.

The *semantic approaches* in this new paradigm are also based on the idea that information changes, and so do the assigned truth-values; and this change needs to be reflected by the formal structure. The *dynamics of information* is the most important concept that arises with this considerations. The fact that the new approaches adopt a dynamic perspective implies that information is not a static concept any more, instead it is in constant motion. What brings the informational entities to move are the *epistemic actions*, that are formalized as operators that change the model. This branch of logics classifies the actions that promote the information to change.

Dynamic Epistemic Logic. The designation of *Dynamic Epistemic Logic* (DEL)[4] includes different approaches, sharing as a common ground Epistemic Logic as their foundation and epistemic actions as its modification. One of the most salient handbooks regarding the dynamics of information is van Ditmarsch et al. (2008). In

[4]The first research that considers the appearance of DEL a change of paradigm is Gochet (2002). DEL as it is considered here is best developed in van Benthem (2011).

there, the authors define very precisely in which sense the concept of 'information' is now interpreted.

> We regard information as something that is relative to a subject who has a certain perspective on the world, called an *agent*, and the kind of information we have in mind is meaningful as a whole, not just loose bits and pieces. This makes us call it *knowledge* and, to a lesser extent, *belief*. This conception of information is due to the fields known as *epistemic* and *doxastic logic* (van Ditmarsch et al. 2008, p. 1. Original italics).

As mentioned above, DEL serves as a designation for different approaches and logics, inspired partly by EL and partly by Belief Revision theory. The different logics and their interactions that have been proposed in the last decades conform a wide range of possibilities. I will mention the most important notions that appear in those proposals and that make DEL constitute a new paradigm.

The most relevant notion is that of *epistemic actions*, understood as the fulfilment of an epistemic act that causes the existing information of the agent to change by loosing information, gaining new pieces of information or transforming it. The first and most simple action that was formalised is the act of communicating new information, captured in **Public Announcement Logic** (PAL) (see Plaza 1989, 2007). I called it 'simple' since this logic formalises an idealisation of the process of communication, i.e. it represents informational changes "per the occurrence of completely trustworthy, truthful announcements" (Baltag and Renne 2016). This logic deals with how the announcement of new information alters the existing one, changing part of what the agent had and creating new information derived from the old one.

Different logics that include different types of information loss or wins and updates have been developed. The variety of epistemic actions correspond to a huge classification that keeps on growing with every new research. Of course, as happened with standard EL, one also finds here the problem of logical omniscience and some actions are designed to prevent the agents from different types of omniscient properties, while other maintain some closure properties and have been applied to Computation, Artificial Intelligence or Economics.

The main feature that defines non-omniscient agents is to have **limited reasoning abilities**. It is not intended that agents cannot reason at all, but to prevent full logical omniscience the agents each logic presents need to have some kind of limiting mechanism that makes their inferential processes not infinite and provides them with just enough rationality for creating new information by themselves, but at the same time prevents them from being able to be informed about everything there is.

In line with this, there have been some very interesting proposals that, based on Awareness Logic and adopting a dynamic perspective, incorporate the actions of *becoming aware* (and also unaware) or *performing a deductive inference*. Both are individual actions (only affecting one agent) that help limit the deductive possibilities of the agent, while reconciling itself with the common sense. These proposals mix different syntactic and semantic structures in order to be able to capture the intended consequences of each action.

To sum up, one could understand the new paradigm of the dynamics of information as an interdisciplinary field of study, where logicians, linguists, computational engineers and philosophers have established as a common ground the concept of 'information change', and constructed a new branch of knowledge. Contemporary advances in the field are constant and new interpretations of the basic concepts come with almost every new proposal that is conceived.

1.2 Epistemology: An Overview and a Choice

This section starts by introducing the concept of *Epistemology* and its historical origins. After describing some of its basic ideas it explores the contemporary picture of Epistemology and reviews some classifications. Then, one specific approach, Awareness Justification Internalism, is highlighted as the chosen view for this research.

Definition of the term 'Epistemology'. The term *Epistemology*[5] has suffered from several semantic changes. Nowadays, Epistemology is the branch of Philosophy that studies knowledge. There is no consensus for a commonly shared definition. Going through the most common dictionaries, one finds the following interpretations: the *Oxford Dictionary* defines it as "[t]he theory of knowledge, especially with regard to its methods, validity, and scope, and the distinction between justified belief and opinion."[6] While the *Merriam-Webster Dictionary* says that it is "the study or a theory of the nature and grounds of knowledge especially with reference to its limits and validity".[7] Turning to specialised literature, Hendricks says that "[t]he systematic and detailed study of knowledge, its criteria of acquisition and limits and models of justification is known as epistemology" (Hendricks 2005, p. 1).

Rivers of ink have already flown regarding the different interpretations of the term and in which sense it is related to all types of knowledge or information. For the present purpose I will stick to the first and simple definition I gave above: *branch of Philosophy that studies knowledge*. As such, there are many different aspects regarding the notion of knowledge that may fall under the epistemological investigations. Of course not every approach to knowledge is suitable of being considered Epistemology, but nowadays this field of study has widen its range in such a way, that aspects like perception, communication, social factors or cognitive processes are also part of it.

Origins of Epistemology. The topic of Epistemology has been one of the main themes in philosophical discussion since its very beginning in Ancient Greece. One could establish its origin in Plato's dialogues. It is said that Plato states that knowledge is equivalent to *justified true belief* (JTB) in the *Meno* and the *Theatetus*. Though this claim is not accurate and those who state it in these terms are accused of anachronism,

[5]Despite of the fact that the topic is very old, the term itself was coined for the first time in Ferrier (1854).

[6]In https://www.lexico.com/definition/epistemology, accessed 01/02/2020.

[7]In https://www.merriam-webster.com/dictionary/epistemology, accessed 01/02/2020.

what may be claimed without falsehood is that in these dialogues, Plato equals 'knowledge' with 'true belief' and adds some additional condition, which might be called 'justification' in modern terms. Attributing modern terms to ancient savants is always complicated and the lack of accuracy is a constant thread. Thus, it should suffice to claim that the origin of Epistemology can be traced back to Plato's dialogues where the thesis that 'knowledge is justified true belief' was originated.

The JTB-view (as Plato's understanding is called) is often used to define traditional Epistemology by itself. The different ways in which 'justification' can be interpreted or understood will give rise to different points of view or branches of Epistemology. The most general classification is made regarding the source of the justification. Thus, on the one hand, there is *Epistemic Internalism*, that claims that the justification for true beliefs happens 'inside' the subject, it is an internal matter where the actual world plays no role at all; and, on the other hand, one finds *Epistemic Externalism*, stating that whatever justifies a true belief must stem from 'outside' the subject, that is, it is the external world that determines the justification. Both, Epistemic Internalism and Externalism, are still being continued in mainstream Epistemology and promote endless arguments and counterarguments, as will be explained below.

Going back to the historical foundations of Epistemology, specifically Epistemic Internalism, the JTB-view was 'adopted', many centuries later, by René Descartes. I risk here being accused guilty of anachronism, since Descartes himself never uses 'justified true belief'. What is clear, is that in Descartes' Rationalism knowledge is all that of which no subject can doubt about its truth and is achieved by means of reasoning. And since *all that* might be everything the subject believes, one could somehow claim that he refers to *justified true belief*. Descartes' Epistemology is seen as one of the first variants of Internalism. The distinction between Internalism and Externalism goes hand in hand with the difference between Rationalism and Empiricism, respectively, contemporary to Descartes' era.

Currently, the view by René Descartes is referred to as *the Cartesian view* and it is assumed to suffer from two big epistemic problems: *scepticism* and *solipsism*.

Scepticism, knowledge is problematic or doubtful, represents Descartes' biggest challenge from the very beginning of his philosophy.[8] As such, Epistemic Internalism is suitable of leading to scepticism. The possibility of knowledge is questioned here and it is quite complicated to find a trustful and truthful source for knowledge in the Cartesian view, since his metaphysical convictions forced him to doubt absolutely every source of knowledge. As a consequence of that, Descartes' theory also falls into *solipsism*, the self is the only thing to know. The ultimate consequence of the Cartesian method, after doubting every source of truth is that only the self can be known, the problem arises when he tries to build new knowledge stemming from the self and doubting every external source. The ways in which Descartes solves this problem and the suitability of it is of no concern to this research, hence I will just close this brief historical review by mentioning that Descartes appeals to God as source of omniscience to help the solipsist subject to gain knowledge again.

---

[8] See, for instance, Chamizo-Domínguez (1984) and Gilson (1950) for a discussion about the origin of Descartes' thoughts and motivations.

Currently, one could say that the Cartesian view of knowledge established that "it is a conceptual truth, that if conditions C justify belief B for subject S, then C logically entails that B is true" (Cohen 1984, p. 280). Being this so, the step to scepticism is straightforward: any conditions C may be established, that make B false.

Epistemic Thought Experiments. One of the most common argumentative course of action used in mainstream Epistemology is based on thought experiments. An epistemic thought experiment being a theoretical and ideal situation where a subject (or more) is put in an idealized, and sometimes implausible, context, where she experiences some type of delusions or misconceptions. Most thought experiments are created for arguing in favour of either Internalism or Externalism.

Probably, one of the most well known thought experiments that was coined in the past is the *Evil Demon Case*, suggested by Descartes, in order to achieve the universal doubt.[9] In contemporary epistemological approaches there has been a remake of this experiment, denoted as the *New Evil Demon Case*.[10] This experiment, similar to the original, consists in a situation where there is an Evil Demon fooling the subject in a sense that every information she obtains by perception (empirical knowledge) seems false, though it is true. With the New Evil Demon Case the internalists prove that the external factors do not play a role in the acquisition of knowledge and, even more, that paying attention to them could lead to false beliefs and misconceptions.

Although it is not exactly a thought experiment, I have to mention at this point the famous *Gettier-Cases*. In Gettier (1963), a paper that is only three pages long, the author states one of the most important challenges to Epistemic Internalism. Gettier presents two plausible cases that show that a subject can have false beliefs that, nevertheless, serve as justifications for true propositions; resulting in scenarios where a subject 'knows' a proposition that is true, but her justifications for it are based on false beliefs that accidentally turned out to validate the true proposition.

On the other side, regarding experiments made up by externalists to prove their point, one could highlight the *Twin Earth Thought Experiment*, presented in Putnam (1973). This experiment is based on the idea that there is a Twin-Earth, where everything looks the same as on Earth, but the chemical composition of the water differs from the water on Earth, hence twin-water is made of XYZ instead of H_2O. Then, he presents a subject, Oscar, that lives on Earth, and Twin-Oscar, that lives on Twin-Earth; and proposes a scenario where the subjects switch worlds, such that now Oscar is on Twin-Earth and Twin-Oscar is on Earth. Both are swimming in a pool and they know that they are swimming in water, but their thoughts will be false, since Oscar thinks he is swimming in H_2O, while swimming in XYZ and vice versa. What this thought experiment tries to prove is that the external world determines the semantic content of our thoughts and hence also the justifications of them.

---

[9]Though the term 'thought experiment' corresponds to the common terminology of Epistemology, some specialists on Descartes would claim that this example is a hypotheses, instead of an experiment.

[10]Stated for the first time in these terms in Cohen (1984).

As should be already clear, neither of those thought experiments are absolute or definitive when it comes to proving one's point. The argumentative style adopted in mainstream Epistemology has resulted in a very long list of literature that not only discusses and rewrites these (and other) thought experiments, but also creates new versions of both Internalism and Externalism.

Contemporary Epistemology. Contemporary Epistemology is such a wide-ranged field that one needs to delimit it, if the term 'Epistemology' is intended to denote something at all. To do so, I will allude to Hendricks, who draws the difference between 'Mainstream and Formal Epistemology' in his homonymous book in the following sense:

> Contemporary epistemological studies are roughly carried out: (1) in a *mainstream* or informal way, using largely conceptual analyses and concentrating on sometimes folksy and sometimes exorbitantly speculative examples or counterexamples, or (2) in a *formal* way, by applying a variety of tools and methods from logic, computability theory or probability theory to the theory of knowledge. The two traditions have unfortunately proceeded largely in isolation from one another (Hendricks 2005, p. ix. Original italics).

'Formal Epistemology' corresponds, in a wide sense, to Epistemic Logic and its developments. As explained in the previous section, EL is a logical representation of knowledge (and doxastic logic, of belief) but it has little to do with mainstream Epistemology. As Hendricks himself points out: "it is a discipline devoted to the logic of knowledge and belief, but is alien to epistemologists and philosophers interested in the theory of knowledge" (Hendricks 2005, p. 81). Therefore, in what follows, I will continue this exposition sticking to mainstream Epistemology and turn now to contemporary approaches.

To reassure what is meant here with 'mainstream Epistemology' and erase every whisper of doubt there might still be, I will quote again Hendricks, who, in a more revealing way, states the following about how mainstream Epistemology is conducted:

> The term 'mainstream epistemology' refers to the *modus operandi* of seeking necessary and sufficient conditions for the possession of knowledge based on the standard definition or some close derivative thereof. [...] It is a dialectical and sometimes even 'diabolical' process that by its very nature balances between the theory of knowledge and skepticism (Hendricks 2005, pp. 14–15. Original italics).

Fortunately, in contemporary Epistemology, there are many authors who claim to have solved this 'scepticism-alert', at least from an internalist point of view. Epistemic Internalism stems from the aforementioned *Cartesian View*, defined by Cohen in terms of the 'conditions C' that justify 'belief B' for 'subject S', such that 'C logically entails B'. This view results very easily in a vicious circle, when C is compatible with B being false; or when a *Gettier-Case* is explained in those terms.

Most forms of contemporary Epistemic Internalism share, thus, a common ground: they impose some *additional restrictions* on conditions C such that the vicious circle can be broken and the true belief is guaranteed.

These restrictions deal with the way in which the notion of *justification* is interpreted and how the subject accesses the information at her disposal. Of course the label 'internalism' is related to an internal source for knowledge; but this does not conflict with the claim that the *grounds for a justification* may come from an external source. Some epistemologists have claimed, that accepting an external source leads directly to some form of Externalism, I prefer to consider it a more intuitive and accurate understanding of the complex process of gaining knowledge, where the creation of knowledge is, indeed, internal to the subject, but the origin of the information known by her, stems from her perception of the world (external source). Hence, Internalism and Externalism, views that started out being complete opposites, represent two elaborated and wide-ranged theoretical frameworks that may share common positions without colliding.

Contemporary Epistemic Internalism. The frontiers between Internalism and Externalism are being blurred in increasingly more approaches. Nevertheless, for the purpose of this research, I am interested in the new approaches to Epistemic Internalism. Thus, I will briefly sum up the contemporary perspectives that fall under the designation of 'Internalism'.

In Pappas (2017) one finds a very extensive review of the different understandings of 'epistemic justification' from both the internalist and externalist point of view. What I want to highlight from this classification are the different useful concepts when contemplating the overall picture. First, the author names *Knowledge Internalism* all those approaches that deal with "knowing or being aware of that on the basis of which one knows" (Pappas 2017, para. 1), as analogous to what justifies the beliefs.

In *Knowledge Internalism*, there are different interpretations. What is common to all of them is the importance that the concept of *accessibility* has. How and to which amount the subject accesses the *knowledge basis* that acts as a justifier will determine the type of *Accessibility or Awareness Knowledge Internalism* that is being proposed. In this classification there is a *strong* and a *weak* version of Awareness Knowledge Internalism, with the former forcing the subject to be aware of something that is indeed in her knowledge basis, and the latter demanding the subject to be aware of everything that is in her knowledge basis.

Since, as explained at the beginning of this section, 'knowledge' is sometimes equated with 'justified belief', there are versions of these same proposals that instead of focusing on 'knowledge', are centred on 'justification'. In this manner, one finds a weak and a strong version of *Awareness Justification Internalism* (and also Externalism, that corresponds to the denial of the internalist proposal). The main difference between the 'Knowledge Internalism' and 'Justification Internalism' being that instead of a 'knowledge basis' that justifies the belief, the latter proposes a *justifier* as the umbrella term for all that information that is suitable of justifying other beliefs that will become knowledge.

While accuracy in the terms is always wanted, I will stop making detours and take both views as analogous accounts for the *Awareness Justification Internalism*. This view is one of the contemporary proposals that overcomes the problem of scepticism, as well as resists most critiques from the *Gettier-Cases*. It does so by adding *awareness* as an additional requirement to the conditions that justify the beliefs of the agent.

Awareness Justification Internalism (AJI). What attracts attention to this type of Internalism is that it imposes what Bergmann calls the *Awareness Requirement*. In his terms:

> *The Awareness Requirement*: *S*'s belief *B* is justified only if (*i*) there is something *X*, that contributes to the justification of *B* –e.g. evidence for *B* or a truth-indicator for *B* or the satisfaction of some necessary condition of *B*'s justification– and (*ii*) *S* is aware (or potentially aware) of *X* (Bergmann 2006, p. 9. Original italics).

Thus, this type of Epistemic Internalism is the view that considers that for a subject to have knowledge, she not only needs to be justified in what she believes, she also needs to be aware (or potentially aware) of whatever made her believe as she does. In BonJour and Sosa (2003), the authors present a thorough review of the different interpretations of this 'Awareness Requirement' and the ways in which the acceptance of a valid external source for justification is incorporated into Epistemic Internalism. What they highlight is the fact that what contributes to the justification, must be "cognitively available to the believer himself, within his cognitive grasp or ken" (BonJour and Sosa 2003, p. 24).

By making the subject aware of these grounds, this Epistemic Internalism manages to escape scepticism and develops into a 'consistent' theory of knowledge. According to Madison (2010), for a coherent version of Internalism, there are some factors that need to be taken into account: (1) the accessibility of the grounds, (2) the adequacy of the grounds, i.e. the grounds that support the beliefs must be relevant to them, and (3) the basing relation, that is, what ties together the beliefs with its justifications. These factors may seem quite obvious, but there are many approaches and interpretations of Internalism that do not place enough value on them. I agree with the author, in the fact that these factors are not only important, but necessary, for a 'logically meaningful' view of Epistemic Internalism. He also states that AJI takes these factors into account.

1.3 Epistemic Logic and Epistemology: Its Disconnection and Awareness as a Bridge

It is worth highlighting the existing disconnection between Epistemic Logic and Epistemology regarding their main tasks. Though there is a thread of hope, when one analyses the latests developments in the new Epistemic Logics that form the new paradigm of dynamics of information, this does not mean that one can state a 'real' connection, yet.

A hint of the current state of affairs can be read in the quote by Hendricks, that ends with: "[t]he two traditions have unfortunately proceeded largely in isolation from one another" (Hendricks 2005, p. ix). This *isolation* or disconnection is, in part, due to the high degree of specialisation that Epistemic Logic has gained in the past decades, including the terminological specialisation, which hinders most epistemologists to keep an updated picture of the latests developments.

The disconnection. Both fields of study are devoted, broadly speaking, to the same main concepts, namely those related to 'knowledge'. As such, the intuitive thought is that the logic of knowledge and the theory of knowledge should work hand in hand and that their new advances are due to a strong collaboration. But nothing could be further from reality.

As partly explained above, Epistemology (in its mainstream developments) is devoted to the different theories of the origin of knowledge, resulting in Epistemic Internalism and Externalism. Epistemology is thus centred on the exposition of arguments and counterarguments either in favour or against one of the existing views, and on proposing new versions of these views that overcome the traditional problem of scepticism, as well as on presenting completely new points of view. On the other hand, Epistemic Logic and its contemporary developments are based on the representation of all types of informational attitudes (including knowledge) and their changes during the communication process (dynamic perspective).

Hence, there are two separated areas, that deal with the same concepts and do not take the other into account. The reasons for this disconnection are diverse and there is no consensus on the factors that determine the rupture between them. From my point of view, the fact that strikes me more plausible is related to the high degree of specialisation of Epistemic Logic, as advanced before.

It is a fact that, since the 1980s and 1990s, Epistemic Logic (in accordance with other logics) has experienced a tremendous growth, thanks to the parallel development of Computer Science and its mutual interest. Most logics have developed into huge specialised fields of study and have become inevitably more and more technical in their methods and manners. And I say 'inevitably' since it is a natural effect of the use of a complex mathematical apparatus. As the logical calculus becomes more complicated, so does the degree of expertise of the logicians increase, for achieving the required level. And, in turn, this implies that most contemporary logicians are specialised in their own field and do not 'waste time' on more philosophical aspects.

The last paragraph could be seen as a pessimist account to the disconnection, since it leaves little space for philosophical discussion and presupposes a false superiority of Logic compared to Epistemology, or Philosophy in general. Nothing could be more wrong than this. Of course Logic is an independent research field, as it has always been, the main difference today being probably that in order to become a good logician, one needs also to manage mathematical calculus to a high degree and preferably also some Computational Engineering. But this does not, and cannot, imply any type of superiority when it comes to comparing it with other studies. The great value of Epistemology is indisputable, such as that of all branches of Philosophy in modern times. I think that it is even more necessary today, than it was a century ago, to keep on doing Philosophy and spreading its importance.

So, the only objective reason for the disconnection is simply that logicians and epistemologists do not share a common ground any more, not even when it comes to their terminology. But is it really like this? Well, in the big picture it certainly is, and that is why the disconnection is so clearly perceived. But as will be shown below, there is still much hope for a new convergence.

Tending bridges During the last two decades, there have been some attempts to bring both fields back together, papers like Hendricks and Symons (2005) and van Benthem (2006) are good representatives of those attempts. The main goal of these papers, and also collections such as Arló-Costa et al. (2016), is to unravel the interconnections that have always been there, but are sometimes difficult to observe.

Hendricks and Symons find the 'bridges' they are looking for in the central notions that define both fields, namely "knowledge, belief and doubt" for mainstream Epistemology, and "learning, information and strategies" for Epistemic Logics. They state that "these two sets of notions are congruent and parallel" (Hendricks and Symons 2005, p. 160). What these authors show in their paper is that though with different technicalities, Epistemology and Epistemic Logic are still dealing with the same problems and solving them in quite similar ways. Their general attitude is based on the idea, that there are several notions that act as a bridge between both fields and that future proposals will learn from both.

Turning now to van Benthem (2006), the author shows a positive view regarding the interconnections of Epistemology and Epistemic Logic. As mentioned before, there is a thread of hope that both areas will merge again when applying the dynamic perspective to information, and this is exactly what van Benthem does in this paper. He reviews most of the central notions from the new paradigm of the studies of language and information (which, in turn, determine the new Epistemic Logics that are being developed today) and concludes that these concepts are not only present in mainstream Epistemology, but some may also stem from there. He is also positive about the future interactions and, to sum up, presents what I like to call a 'virtuous circle', where the results of a collaboration can only be those of a mutual benefit.

Parallel development of Epistemology and Epistemic Logic. It is clear from the foregoing, that Epistemology and Epistemic Logic represent two different research areas with its own terminology and methods, and that there is a common historical background to both fields that will help me bring them back together. For doing so, it is very helpful to present the development of both fields in parallel, showing thus the interconnections this book will analyse.

Regarding the tradition, or historical origin, Epistemology stems from the dispute between Internalism and Externalism regarding the justification of the 'true belief', while Epistemic Logic has its origin in the extension of Modal Logic that applies its structure to the knowledge of epistemic agents. Both subjects have been challenged with a problem that threatened their very existence. Epistemology, specifically Epistemic Internalism, suffered from the problem of Scepticism, which for years has been the burden every internalist had to solve. On the other hand, Epistemic Logic was challenged with the problem of Logical Omniscience.

The respective solutions to these challenges share the same structure, namely to impose some kind of restriction to the main concept that is being questioned. In the case of Epistemic Internalism it is the concept of *justification* that will require some additional restrictions, while in Epistemic Logic it will be the *knowledge* (in most cases the explicit version) that has to be restricted due to some additional element.

As already mentioned regarding both areas, there are many different solutions, but the two I will consider along this research are the view of Awareness Justification Internalism (AJI) for Epistemology and Awareness Logic (AL) for Epistemic Logic. To continue this parallel presentation the concept that is added to both fields as the new element that solves the respective problem is the concept of *awareness*. In AJI in the form of the Awareness Requirement, that limits the justification, and in AL as the awareness operator, that serves as a filter on knowledge. I state that for the purpose of the present research the concept of awareness is the appropriate tool. To which extend this is the case will be revealed in what follows.

The concept of Epistemic Awareness. So far, every appearance of the term 'awareness' has been made in its 'epistemic' understanding, but before continuing an argument in favour of this concept, I think that a more thorough specification of the notion is needed. 'Awareness' is a very polysemic term, sometimes equated with 'consciousness' and others applied to a long list of perceptual features of human beings. This ambiguity brings me to the need of clarification.

I connect 'awareness' with an epistemic understanding of 'perception' (the information one perceives)[11], and 'consciousness' with all types of perceptions, including their moral and psychological implications. That is, the thoughts about 'reasons for actions' or 'the right thing to do' may be part of what 'consciousness' alludes to in a general picture, while they should never be part of a subject's awareness. Of course, in more informal contexts both terms are used as synonymous and it is common to read phrases like 'raise of awareness' or 'being conscious that something is the case', but with the former distinction and the definitions below I intend to overcome this confusions and establish a common terminological background for this research.

The 'awareness' I will be referring to is always epistemic, that is, it accounts for the fact of *realizing one's own information*. Attending the standard definitions, we find in the *Oxford Dictionary*[12] 'awareness' as "[k]nowledge or perception of a situation or fact", and in the *Merriam-Webster Dictionary*,[13] "the quality or state of being aware: knowledge and understanding that something is happening or exists."

Though neither of these definitions allude to the verb 'to realize' I mentioned in my own definition, it is somehow implied, since the definitions of this verb include the notion of 'awareness'. The first entry of 'to realize' in the *Oxford Dictionary*[14] says "[b]ecome fully aware of (something) as a fact; understand clearly"; and the third entry of the *Merriam-Webster Dictionary*[15] defines it as "to conceive vividly as real: be fully aware of".

[11] In many places, like MacMillan (2012), this understanding is referred to as 'conceptual awareness'.

[12] In https://www.lexico.com/definition/awareness, accessed 01/12/2020.

[13] In https://www.merriam-webster.com/dictionary/awareness, accessed 01/12/2020.

[14] In https://www.lexico.com/definition/realize, accessed 01/12/2020.

[15] In https://www.merriam-webster.com/dictionary/realize, accessed 01/12/2020.

As such, I stick to my simple definition of 'awareness' as *realizing one's own information*. Here, I have to highlight two aspects: (*i*) with 'one's own' I refer to everything the agent is able to entertain using her rational means, not only what she observes, but also what she infers from it; and (*ii*) the perceptions relevant for 'awareness' are always understood as information, in the wide-ranged meaning of this term, introduced in Sect. 1.1. As such, this includes also higher order information, i.e. that she might be aware of her own awareness or aware that someone else is aware of something (in cases of a multi-agent scenario).

Awareness in Epistemic Internalism. As already explained in Sect. 1.2 the concept of 'awareness' is introduced in a family of internalist views that allude to the 'access' of the subject to the justifying information. This type of Internalism is named *Accessibility or Awareness Justification Internalism* (AJI) (also known as 'Awareness Internalism').

My support for AJI is based on several reasons: *first*, because I think that all the other alternative views, whether internalist or externalist, fail to capture what contemporary cognitive science has already explored, namely that the role of awareness is intrinsic to the process of forming knowledge and belief. *Second*, because this view is favourable for serving as a correlate to some Epistemic Logics, or, to put it another way, because this internalist view can serve as a foundation for some Epistemic Logics (as will be made clear in the following chapters). *Third*, and this might be the less substantiated reason, because I think that AJI captures the common sense intuitions behind the knowledge-forming process. Of course, when one alludes to 'common sense intuitions', it is hard to establish an analytically structured argument, since the 'common sense' is a very subjective and also individually oriented point of view that pursues to be held by a large group of individuals that belong to the same society. Being this so, I still maintain that AJI captures what is held by common sense, at least, for the logically oriented and analytically formed philosophers of the twenty-first century.

The importance of an *Awareness Requirement* in Epistemic Internalism becomes more relevant when one attends the implications it entails. Requiring the subject to be (potentially) aware of the grounds on which she justifies her belief means that the subject is rational and is capable of reflecting upon her own beliefs. This feature has been criticised by some epistemologists, arguing that it "seems to suggest an over-intellectualized and deliberative picture of our belief-forming activities. [...] But many of our beliefs are not formed in such a deliberative manner" (Vahid 2010, p. 150).

I agree with the author in the fact that not every belief a subject entertains is fruit of a completely rational process, but those upon which she builds a justification for her knowledge, i.e. those beliefs that she will consider true and will be able to justify, should, of course, be rational. If not the process of belief-forming itself, then at least she should be able to trace back the grounds for holding it, and in that, awareness is crucial.

From all the above reasons I dare to say that Awareness Justification Internalism is *the* view that best suits an epistemic agent with limited reasoning abilities. I would like to recall here that my main goal is to establish a meaningful theoretical framework that can serve as a foundation for the new developments of Awareness Logic with epistemic actions. And this, in turn, implies that I will use Epistemology, concretely Awareness Justification Internalism, as a guide for this theoretical framework, but I do not pretend to *make* Epistemology, meaning that I do not want to enter the countless debates that epistemologists remarkably perform in a structured and argumentative manner. Thus, at this point, I will stop advocating why I choose this internalist view and move on to explain why I choose awareness as *the* solution to logical omniscience from a logical point of view.

Awareness as a solution to logical omniscience. In Section 1.1 I reviewed the different solutions to the problem of logical omniscience that Fagin et al. mention in their book and already anticipated that *Awareness Logic* strikes me the best solution. I shall now present the reasons that lead me to this statement.

All the mentioned solutions (and probably a long list with more) can be equally valid or appropriate from a strictly logical or computational point of view. If the only aim that a solution to the problem pursues is to limit the agent's reasoning abilities, independent of how this agent is being represented, then of course all the solutions fit their goal. But my aim here is not strictly logical and hence, there are solutions that though computationally adequate, make use of a complex mathematical apparatus or logical structures and thereby might loose their connection to the reality of human knowledge.

The non-standard structures, the impossible worlds approach and the purely syntactic approaches seem very useful for capturing some concrete aspects of human knowledge, but fail to capture a bigger picture. I want to emphasise that, though equally valid, not all of them manage to fit into a trustworthy representation of human knowledge.

I also want to highlight that my humble intention with this research is not only to establish a conceptual schema that suits a new version of Awareness Logic as its foundation, but also to make this new version a very simple one. This is the reason why I rule out all the non-standard and more complex solutions, as well as the multi-agent proposals that there already are. I prefer to stick with a logic that is simple and accessible to a wider audience, rather then engage myself with more complicated structures, whose intrinsic features would not be easy to relate with most theoretical notions. Being this so, this research includes a formal model for representing Explicit Aware Knowledge (contained in Chapter 6) that can also be seen as another possible solution.

Regarding the concept of 'awareness', in (Fagin and Halpern 1988) the authors present the first 'version' of Awareness Logic, namely "A Logic of General Awareness". They propose a very vague concept and make their intentions clear in the following quote:

> We have not yet discussed exactly what 'awareness' really is, and indeed, we do not intend to do so here at all! The precise interpretation we give to the notion of awareness will depend on the intended application of the logic (Fagin and Halpern 1988, p. 53).

This claim could be seen as an example of the standard methodology from contemporary logics: a concept, after imposing restrictions, can be whatever suits best the intended outcome. In contrast to this, I want to establish what the concept *is* and then adapt the logical interpretation to it.

As such, starting from the concept of 'awareness' as defined before, and having in mind its relevance from an epistemological point of view, I am in a position to claim that awareness is the appropriate tool for solving the problem of logical omniscience, i.e. for representing explicit knowledge of an agent with limited reasoning abilities, as well as for the epistemological understanding of knowledge captured by Awareness Justification Internalism. This being so, it should be established that the concept of *awareness* is the bridge that will allow me to re-connect the fields of Epistemology and Epistemic Logic.

References

Alchourrón, C. E., Gärdenfors, P., & Makinson, D. (1985). On the logic of theory change: Partial meet contraction and revision functions. *The Journal of Symbolic Logic*, *50*(2), 510–530.

Arló-Costa, H., Hendricks, V. F., van Benthem, J., (Eds.). (2016). *Readings in Formal Epistemology: Sourcebook*. Dordrecht: Springer.

Baltag, A., & Renne, B. (2016). Dynamic epistemic logic. In Zalta, E. N. (Eds.), *The Stanford Encyclopedia of Philosophy*. Metaphysics Research Lab, Stanford University, winter 2016 edition.

Bergmann, M. (2006). *Justification Without Awareness: A Defense of Epistemic Externalism*. Oxford: Oxford University Press.

BonJour, L., & Sosa, E. (2003). *Epistemic Justification: Internalism Vs Externalism, Foundations Vs Virtues*. Hoboken: Wiley-Blackwell.

Chamizo-Domínguez, P. J. (1984). *La doctrina de la verdad en Michel de Montaigne*. Málaga: Universidad de Málaga.

Cohen, S. (1984). Justification and truth. *Philosophical Studies*, *46*(3), 279–95.

Fagin, R., Halpern, J., Moses, Y., & Vardi, M. Y. (1995). *Reasoning About Knowledge*. Cambridge, MA: The MIT Press.

Fagin, R., & Halpern, J. Y. (1988). Belief, awareness, and limited reasoning. *Artificial Intelligence*, *34*(1), 39–76.

Ferrier, J. F. (1854). *Institutes of Metaphysic. The Theory of Knowing and Being*. Edinburgh and London: William Blackwood and Sons.

Gärdenfors, P. (Ed.). (1992). *Belief Revision*. New York: Cambridge University Press.

Gettier, E. (1963). Is justified true belief knowledge? *Analysis*, *23*(6), 121–123.

Gilson, É. (1950). *The Unity of Philosophical Experience*. New York: Charles Scribner's Sons.

Gochet, P. (2002). The dynamic turn in century logic. *Synthese*, *130*, 175–184.

Hawke, P., Özgün, A., & Berto, F. (2019). The fundamental problem of logical omniscience. *Journal of Philosophical Logic*, *49*(4), 727–766.

Hendricks, V. F. (2005). *Mainstream and Formal Epistemology*. Cambridge: Cambridge University Press.

Hendricks, V. F., & Symons, J. (2005). Where is the bridge? Epistemology and epistemic logic. *Philosophical Studies*, *128*(1), 137–167.

Hintikka, J. (1962). *Knowledge and Belief*. Ithaca: Cornell U.P.

Hintikka, J. (1975). Impossible possible worlds vindicated. *Journal of Philosophical Logic*, *4*(4), 475–484.

MacMillan, I. D. (2012). *Internalism, Externalism, and Epistemic Source Circularity*. Ph.D. thesis, University of Iowa.
Madison, B. J. C. (2010). Epistemic internalism. *Philosophy Compass*, *5*(10), 840–853.
Pappas, G. (2017). Internalist versus externalist conceptions of epistemic justification. In Zalta, E. N. (Ed.), *The Stanford Encyclopedia of Philosophy*. Metaphysics Research Lab, Stanford University, fall 2017 edition.
Plaza, J. (2007). Logics of public communications. *Synthese*, *158*(2), 165–179.
Plaza, J. A. (1989). Logics of public communications. In Emrich, M. L., Pfeifer, M. S., Hadzikadic, M., & Ras, Z. W., editors, *Proceedings of the 4th International Symposium on Methodologies for Intelligent Systems* (pp. 201–216). Tennessee: ORNL.
Putnam, H. (1973). Meaning and reference. *Journal of Philosophy*, *70*(19), 699–711.
Rantala, V. (1975). Urn models: A new kind of non-standard model for first-order logic. *Journal of Philosophical Logic*, *4*(4), 455–474.
Vahid, H. (2010). Externalism/Internalism. In S. Bernecker & D. Pritchard (Eds.), *The Routledge Companion to Epistemology* (pp. 144–155). Abingdon: Routledge.
van Benthem, J. (2006). Epistemic logic and epistemology: The state of their affairs. *Philosophical Studies*, *128*(1), 49–76.
van Benthem, J. (2011). *Logical Dynamics of Information and Interaction*. New York: Cambridge University Press.
van Ditmarsch, H., van der Hoek, W., & Kooi, B. (2008). *Dynamic Epistemic Logic*. Dordrecht: Springer.
Velázquez-Quesada, F. R. (2011). *Small steps in dynamics of information*. Ph.D. thesis, Institute for Logic, Language and Computation, Universiteit van Amsterdam.
von Wright, G. H. (1951). *An Essay in Modal Logic*. Amsterdam: North-Holland.
Williams, M.-A., & Rott, H. (Eds.). (2001). *Frontiers in Belief Revision. Number 22 in Applied Logic Series*. Dordrecht: Kluwer Academic Publishers.

Chapter 2
Epistemic Awareness as the Main Concept

Abstract Having defined the notion of 'awareness' as *realizing one's own information*, this chapter aims to explore the applications of this concept to both Epistemology and Epistemic Logic. Distinguishing between *awareness of* and *awareness that* will result in two types of awareness, that classify the types of information the agent bears in mind. Based on this distinction a new interpretation of the concept of awareness will be provided and a different understanding of the notion of 'explicit knowledge' will be defined, given rise to the concept of *Explicit Aware Knowledge*, core component of this research.

Keywords Awareness · Awareness requirement · Epistemic action · Change in awareness · Explicit knowledge

2.1 Awareness-Of and Awareness-That: A Key Distinction

Even using 'awareness' always in its epistemic interpretation, its semantic field is still very wide-ranged and a further and very relevant distinction is still needed, namely, the difference between *awareness of* and *awareness that*.

Awareness of and awareness that. So far, every appearance of the term 'awareness' was followed indistinctly by the preposition 'of' or the conjunction 'that' and this might look like a harmless distinction, but there is a big theoretical difference behind it. I want to draw attention to the philosopher Fred Dretske who in Dretske (1993) pointed out this important distinction in the use of 'awareness'.

Though devoted to the Philosophy of Perception, his paper manages to convey this difference in a comprehensible manner. He distinguishes 'awareness of things' (awareness of X) from 'awareness of facts' (awareness that P). In general, the basic intuition behind this distinction (leaving out the nuances that underlie his studies on perception) can be summed up with the following quote:

> Things are neither true nor false though, in the case of events, states of affairs, and conditions, we sometimes speak of them as what makes a statement true. Facts are what we express in making true statements about things. We describe our awareness of facts by using a factive complement, a that-clause, after the verb; we describe our awareness of things by using a (concrete) noun or noun phrase as direct object of the verb. (Dretske 1993, p. 264).

C. Fernández-Fernández, *Awareness in Logic and Epistemology*,
Logic, Epistemology, and the Unity of Science 52,
https://doi.org/10.1007/978-3-030-69606-1_2

Based on his ideas, one could say (in terms more related to logic) that *awareness of facts* stands for the awareness of those thoughts (or believed propositions) the agent can form; while *awareness of things* would be somehow the very fact of perceiving and forming her mental content.[1]

Albeit not directly comparable, one could state that awareness of facts is more complex[2] than awareness of things and represents a smaller range of all the information that could belong to the agent's awareness. But taking the awareness as a whole is of no interest, since what this distinction hints towards is a difference that results in two types of awareness: one formed by all kinds of perceptions and the other comprising 'facts', i.e. complex formulas formed with the perceived concepts. Dretske states that "[i]f S is aware that x is F, then S has the concept F and uses (applies) it in his awareness of x" Dretske (1993), p. 265, relating this way both types of awareness.

The main difference in this distinction lies in the object of the awareness of the agent, i.e. depending on the type of information the agent is bearing in mind, it will be one or the other type of awareness. But from this it follows directly that there *are* two types of awareness with distinct features.

Hence, all types of perceived information will be captured by the awareness of things. On the basis of that, those pieces of information that become a thought or a fact (based on something else, commonly its truth) will be captured by the awareness of facts. I could re-express the distinction by saying that every conceivable information is suitable of being part of the awareness of things; while only the information from which one can form a proposition that will have a truth value, will belong to the awareness of facts.

The terminology 'of things' and 'of facts' may lead to misunderstandings, since the latter is used with the conjunction 'that'. Hence, I will use the expressions *'awareness of'* and *'awareness that'* for my interpretation of 'awareness of things' and 'awareness of facts' respectively. The use of the preposition and the conjunction already distinguishes the type of semantic content the awareness will contain. The main distinction (for the logical purpose) lays in the fact that the content of *awareness of* does not need to have a truth value. One can be aware of objects, of doubts, of beliefs, etc., while the content of *awareness that* intrinsically is understood to be true. An agent can be aware of anything, but if she is aware that φ is the case, than she will know φ to be true. Dretske himself points to this thought referring to a common epistemological assumption: "[i]t is conventional in epistemology to assume that when perceptual verbs take factive nominals as complements, what is being described is not just belief but knowledge" (Dretske 1993, p. 265).

The distinction in Epistemology. Based on this distinction the following question is mandatory: which type of 'awareness' does Awareness Justification Internalism con-

[1] This sense of 'awareness of things' could be related to Quine's 'observation sentences', which he considers opposed to 'propositions' (see Quine 1970 for further details).

[2] The adjective "complex" has to be understood here in its logical meaning, i.e. the logical complexity of formulas. As such, awareness of things will be formed exclusively by atomic propositions, while awareness of facts will deal with more 'complex' formulas.

sider? In other words, which 'awareness' is the *Awareness Requirement* in Bergmann (2006) referring to? The answer is not straightforward, since most proponents of this internalist view do not maintain a 'coherence in the terms' so to say, i.e. at some parts of their researches they employ 'awareness of' and at other parts, they use 'awareness that'.[3] My intention here is not so much to find out which awareness they were mentioning, but rather to establish, based on my understanding of the two types, what role they play in the epistemological view I will support.

Should the epistemologists who favour Awareness Justification Internalism (AJI) be referring to *awareness-that*, this would imply that the agent needs to be aware that the grounds for justifying her beliefs are the case, i.e. she is aware that some information is true and this justifies her true beliefs. This in turn would mean that what she uses as 'justifiers' is already some type of 'knowledge' since she would be sure about its truth. This interpretation could lead to an infinite regress of which many epistemologists have written, regarding Awareness Justification Internalism, but not alluding directly to this distinction.

Hence, the awareness this Internalism renders should not be awareness-that, since it will first, lead to a vicious circle where also every information the agent is aware that it is the case would need to be justified on the basis of some other information she should be aware that it is the case; and second, since knowledge would then be justified true belief on the basis of some other knowledge, namely an aware-that knowledge.

Henceforth, the only option left is that AJI refers to *awareness of* and I do agree with this interpretation. As mentioned before, the agent will be aware of every conceivable information, this could be equivalent to 'everything she is entertaining at this very moment'. As such, one could conclude that if she is entertaining some piece of information, she will be able not only to access it, but also to employ it as the justifying information for her true beliefs, permitting thus the creation of knowledge.

But again, this view might be criticised and accused of another vicious circle, since if the agent is aware of some information, of which she does not consider its truth, then she might also be aware of false information. If this information is intended to act as a justifier for her true beliefs, then cases like the ones in Gettier (1963) may occur very easily. To prevent this, one has to adopt a type of foundationalist view that states that there has to be one basic information whose justification does not depend on other informations (which would need to be justified as well).[4]

---

[3]See, for example, the following quote, where the author is defining his internalist view: "Epistemic internalism, as I am construing it, holds that conscious *awareness of* one's justificatory grounds and *awareness that* those grounds contribute to the justification of one's belief (or the possibility of such awareness upon reflection), is a necessary condition of epistemic justification obtaining" (Madison 2009, p. 175, my italics.).

[4]In Chap. 4, devoted to the concept of justification, this problem will be analysed and 'solved' due to a dynamic understanding of the concept.

Despite the mentioned possibility of criticism, I maintain that the 'awareness' that is at stake in Awareness Justification Internalism is always 'awareness of'. It is also reasonable to assume this, taken into account that 'awareness of' is nothing but another way of referring to the perceptions and it is precisely this notion that is employed in most epistemological approaches. It will be the perceived information that will determine if a subject is able to justify her belief or not.

2.2 Awareness Logic and Acts of Becoming Aware

In Sects. 1.1 and 1.3 I presented the main concepts from Awareness Logic (AL) and explained why I consider it the best solution for the problem of logical omniscience, from a theoretical point of view. In the present section I will carry out an analysis of the concepts from AL and also present other logical systems that incorporate 'awareness' into its structure from the point of view of the dynamics of information. The main goal is to create an overview of the different ways in which both awareness-of and awareness-that have been used in several logical systems.

Awareness in Logic of General Awareness. Regarding the structure of AL there is not much more to say. The important feature to highlight from the proposal presented in Fagin and Halpern (1988), and revised in Fagin et al. (1995) is that they take the system from standard Epistemic Logic and add awareness as a syntactic operator. The awareness they incorporate corresponds to *awareness of*, since they consider it as an arbitrary list of formulas and hence, the agent can be aware of any formula, regardless of its truth value. Together with the incorporation of this operator, they add a new element to the semantic model (see Sect. 1.1), namely the *awareness function*, defined individually for every agent as a kind of 'bag' where every information the agent is aware-of may fit in.

Since the authors apply the concept of 'awareness of' to any formula, they clearly want to leave the notion undetermined. They even highlight this indeterminacy as a positive aspect of their system by saying the following:

> The notion of awareness we use in this approach is open to a number of interpretations. One of them is that an agent is aware of a formula if he can compute a given situation within a certain time or space bound. (Fagin and Halpern 1988, p. 41).

That 'being aware of' can be equated with 'computing in a given time or space bound' shows the versatility of the concept, whose initial requirements are left blank. Of course the authors propose how to introduce several closure properties, but it was not their original intention. They also state at the conclusions of their paper the following thought:

> We consider the flexibility of this approach to be a point in its favor. Nevertheless, it is clear that further research needs to be done in order to find useful and natural awareness functions. (Fagin and Halpern 1988, p. 69).

As mentioned earlier, this one and the previous quote show how the authors wilfully presented a concept 'without properties'. From a strictly logical point of view, this strategy seems not only appropriate but is also common to most logical proposals that incorporate a new concept. In fact, introducing a notion such as 'awareness' in a logical system and providing from the beginning some very specific properties would probably not have been a wise decision, since what the first version of Awareness Logic intended, was to provide a new framework where the concept of 'awareness' serves as a filter for selecting the explicit knowledge from the implicit one.

Although there is nothing to object from the logical point of view, I want to mention that from a theoretical or philosophical point of view this strategy, common also to other proposals, widens the breach that divides formal logic and philosophy of logic. With this I mean that by not even mentioning the theoretical foundations of the key concepts that underlie a new logical system, these new systems are made of 'orphan' concepts suitable of being 'adopted' by any possible theory or different logical system.

Awareness in Dynamic Epistemic Logic. The field of Dynamic Epistemic Logic is wide-ranged and almost every proposal is suitable for incorporating the notion of awareness into its system. As such, one finds a huge variety of awareness implementations regarding all the logical systems there are. Most of these works incorporate awareness as part of the semantic model and then define epistemic actions that either increase or decrease the set of information contained by the awareness.

As a general overview, one could highlight the fifth chapter of van Benthem (2011), entitled "Dynamics of inference and awareness"; the special edition (Ågotnes and Alechina 2014) whose title is "Formal Models of Awareness", which includes different contributions that cover not only the notion of 'awareness' but also several epistemic actions of awareness gain or loss (unawareness); and the third chapter of van Ditmarsch et al. (2015) devoted exclusively to "Awareness".

Works like van Ditmarsch and French (2009) and van Ditmarsch and French (2011), for example, incorporate changes in awareness applied to atomic propositions and to other agents in the group, as well as the notion of "awareness bisimulation" for defining the relations between different models and the informational changes. Other works, like van Ditmarsch et al. (2013), continue with the line of research from the previous ones and incorporate a new type of knowledge, namely the "speculative knowledge". This is to show that there are indeed many different interpretations and applications of the concept of "awareness change".

Though there are numerous works on the topic, I am interested in those that draw a clear distinction between awareness-of and awareness-that. Therefore, in what follows, I will focus only on two papers, which incorporate this distinction in a very clear and explicit sense.[5]

The paper "The dynamics of awareness" (van Benthem and Velázquez-Quesada 2010) is one of the most relevant ones when it comes to incorporating the notions of *awareness of* and *awareness that* to the structure of DEL. As in the case of Fagin and Halpern (1988), they proceed in the standard way and do not mention explicitly

[5]For an updated bibliographic review on Awareness Logic see Fernández-Fernández (2019).

the theoretical background. Nevertheless, it is possible to read between the lines and unravel their concrete understanding of both types of awareness.

On the introductory pages of the paper they employ the words "explicit 'aware-that' knowledge" (van Benthem and Velázquez-Quesada 2010, p. 6), which already shows in which sense they will understand awareness-that. Namely, as equivalent to the 'real' explicit knowledge (taken from the original Awareness Logic). When defining formally *awareness of* they clarify its meaning in the following terms:

> Notice that this is a matter of attention, and does not imply any attitude pro or con: the agent may believe φ, but also reject it. Stated in other, but related terms, 'awareness of' does not imply 'awareness that'. (van Benthem and Velázquez-Quesada 2010, p. 7).

The last sentence is the key of their understanding. Namely, that considering a given φ does not bear with it the knowledge of this formula. Thus, they separate clearly, as did Fagin and Halpern, the awareness-of from the explicit knowledge (in their case, also the awareness-that).

When it comes to *awareness that* the authors do not explicitly mention it, since they equate it with "explicit knowledge as a defined notion", using their explicit knowledge operator, $\mathrm{Ex}\varphi$. At one point, just after defining explicit knowledge as the implicit knowledge of both the formula φ and being aware of φ ($\mathrm{Ex}\varphi = \mathrm{Im}(\varphi \wedge A\varphi)$) they write "[...] explicit knowledge ("awareness that")" (van Benthem and Velázquez-Quesada 2010, p. 9).

From the above information, one deduces that they employ awareness-of as a concept analogous to 'entertaining some information' and awareness-that as 'having explicit knowledge' of it. It is also interesting to highlight that the definition of explicit knowledge differs from the original one from Fagin and Halpern. In Awareness Logic explicit knowledge is equivalent to the conjunction of both having implicit knowledge and being aware-of ($\mathrm{Ex}\varphi = \mathrm{Im}\varphi \wedge A\varphi$); while in their system, explicit knowledge is the implicit knowledge of both the formula and the awareness of this formula ($\mathrm{Ex}\varphi = \mathrm{Im}(\varphi \wedge A\varphi)$). They justify their preference over the original one alluding to the properties of positive and negative introspection that follow from their definition (see van Benthem and Velázquez-Quesada 2010, p. 9) and also (Velázquez-Quesada 2011, pp. 58–59).

This preference brings back some kind of logical omniscience, since they equate explicitly knowing something with having implicit knowledge of it and implicit knowledge of being aware of it. Recall that implicit knowledge is an idealized notion from standard Epistemic Logic and its properties lead directly to the problem of logical omniscience. Of course, their option is related to other interesting properties they wish their agents to possess, but from a theoretical point of view, these agents do not resemble human beings with limited reasoning abilities. In their system of epistemic dynamic actions, the actions provide the way out of omniscience, since they are able to model the changes in information, but taken as a whole it is still an idealized representation.

Regarding the logical *properties* that could be imposed on the awareness sets of the agents, they mention two of them in the following quote about the epistemic actions and their properties. Though what van Benthem and Velázquez-Quesada want to highlight is precisely that their epistemic actions of 'consider' and 'drop' do not maintain whatever properties are assigned to the awareness sets, a feature that is very important and makes their proposal so interesting, I will take the properties they mention as a guideline for comparing the different methodological differences between Epistemology and Epistemic Logic.[6]

> It is also worthwhile to notice how closure properties one might impose on the A-sets (cf. the already mentioned closure under commutation for conjunction and disjunction, or being generated by some subset of atomic propositions, as in cf. Fagin and Halpern (1988)) are not preserved by the operations *consider* and *drop*. Indeed, we think these properties over-idealize the content over agents' awareness (van Benthem and Velázquez-Quesada 2010, p. 12. Original Italics).

That (i) with every information the agent has at her disposal, i.e. her awareness set, she can form complex formulas where the 'order of appearance' of the information plays no role (commutation of conjunction and disjunction); and that (ii) every possible information she might be aware-of needs to be part of a given subset of all the information there is, meaning that her awareness set, what she entertains, is restricted to a specific "subset of atomic propositions", represent properties that do not always suit the description of human knowledge.

Again, as already mentioned, these two properties stem from a purely logical reasoning and thus seem 'logically' useful and meaningful, but when this considerations are extrapolated to the level of human knowledge or human communication the gap becomes clear. I will analyse in more detail the two properties that van Benthem and Velázquez-Quesada mention in the above quote: first asking why this property should be applied to human subjects and then unravelling what this would mean from both a logical and a theoretical perspective.

(i) Why should the order of appearance of conjunctions and disjunctions matter? In a daily communication process between, say, two persons, communicating $\varphi \wedge \psi$ or $\psi \wedge \varphi$ should not alter the content of what is being communicated. From a logical, or preferably, computational perspective, there can be several reasons why the order of conjunctions and disjunctions might play a distinct role. For example, if a program is supposed to compute the information in a specific order, because each proposition causes a preset action to be executed, then the order of conjuncts does matter. But if the conjuncts only carry a semantic content, as they do in an ordinary communication process, then this property is not a realistic one.

On the other hand, *(ii) why should the totality of the perceived information always be a subset of what the agent could be aware of?*

(a) If one looks at the subset as what she is entertaining at this very moment, then awareness-of is understood as the 'focus of her attention', similar to a flashlight that is illuminating one specific surface; but
(b) if one interprets what the agent is aware-of as not only the information she might be perceiving at this very moment, but also what she might have perceived in the past, then no subset of the information is needed for defining her awareness set.

---

[6]Both properties stem from a longer list of properties that can be found in the seminal work (Fagin and Halpern 1988, pp. 53–57).

Which of these interpretations is the right one is a not answerable question. When it comes to what might define best the agent's awareness-of, a definition like (a) has a logical structure, it is inspired in the formation of the awareness set seen as a mathematical set, a collection of (atomic) propositions the agent is considering at a specific moment. As such the different complex formulas the agent may know explicitly (or be aware-that) must be formed by the atomic propositions she has at her awareness set, being these atomic propositions the *language* she can use at this stage.

A definition like (b) stems from an epistemological background, in the sense that for the agent to be aware-of something, there is no limit or restriction, neither in the past nor in the present time. It is commonly assumed that the subject might be able to access all information she has once perceived, hence being able to use this information as a justification for her 'reasons to know'. A primitive understanding of the Awareness Justification Internalism could lead to this interpretation, where the subject could use as the 'grounds for justifying her belief' every information she was once (or is) aware-of. But a more meditated version of this type of Internalism would probably prefer the interpretation given in (a), since the agent will only be able to justify her beliefs with the information she has at her disposal at a given moment.

The above reflections show several interesting factors regarding the methodological differences there are between Epistemic Logics and Epistemology. They are nothing but the tip of the iceberg and my intention is precisely to unravel these details that bring with them a much bigger topic and constitute examples of the existing gap. I will now turn my attention to another relevant paper that, in contrast to the previous one, does explicitly formalize both awareness-of and awareness-that.

The paper "Syntactic awareness in logical dynamics" (Grossi and Velázquez-Quesada 2015)[7] is devoted to a "syntax-based" understanding of awareness and its connections to DEL. I will not analyse the whole system, but I shall proceed in a similar ways as I did with the previous one. This is one of the most relevant papers regarding my purpose with the EAK-Schema, since the authors state that "[...] the paper develops the first approach that combines both notions of awareness, 'awareness of' and 'awareness that', thus allowing a more realistic representation of non-ideal agents [...]" (Grossi and Velázquez-Quesada 2015, p. 4072). The distinction they draw coincides in many aspects with the EAK-Schema proposal. They state it in the following terms:

> To draw a general distinction *awareness of* concerns the propositions (of a situation or possible world) an agent has available or can resort to in order to express what she knows or doesn't know. On the other hand, *awareness that* concerns the propositions an agent acknowledges as true, possibly as a result of deductive reasoning (e.g., the application of modus ponens to what is currently acknowledged) or after communication with an infallible source (an announcement) or nature (an observation). (Grossi and Velázquez-Quesada 2015, p. 4074. Original italics).

In fact, the way in which they describe both types of awareness at this point of their paper, is equivalent to my understanding. The differences will arise when

[7] Which includes the work from a previous paper: Grossi and Velázquez-Quesada (2009).

implementing these concepts in the logical system they propose and analysing its implications. The notion of 'explicit knowledge' I will defend is a combination of both *awareness of* and *awareness that*. And this notion might seem as taken from this paper, since the authors say that:

> Thus, in order for an agent to know explicitly a given φ, it is not enough for her to know it implicitly (the standard definition of knowledge in *EL*): she also needs to be aware of it (as required in, e g., Fagin and Halpern 1988) and also be aware that it is the case (as required in, e.g., Duc 1997; Jago 2009). (Grossi and Velázquez-Quesada 2015, p. 4074. Original italics)

But the main difference regarding the EAK-Schema will be that the authors still maintain, in some revised sense, the structure of Awareness Logic by insisting that 'explicit knowledge' is formed by the implicit knowledge the agent is aware of and that (in their case). The main point is to decide which concept of 'knowledge' (implicit or explicit) acts as the primary notion.

Regarding the concepts of *awareness of* and *awareness that* which are the point of the present chapter, I shall recap how they implement these notions in their system. For the concept of 'awareness of' they incorporate a new syntactic operator which creates "availability formulas", forming thus the set of formulas the agent is aware of. Regarding 'awareness that' they use the same mechanism, incorporating an operator that creates "acknowledgement formulas".

The awareness-of set is formed by atomic propositions and its semantics is given by the "propositional availability function" that assigns a list of atoms to a given agent for every possible world. This property (already mentioned above) makes this set closed under subformulas and superformulas, provoking that if an agent is aware of φ and $\varphi := p \wedge q$, then the agent is automatically aware of p and q as well as aware of $p \vee q$ or $p \rightarrow q$, for example. It is nevertheless a reasonable feature that being the *awareness of* similar to the 'language' the agent has at her disposal (or the set of everything she entertains at a given moment) she will be able to form the different combinations that the Boolean operators permit with the atomic propositions she entertains. As such, the syntactic interpretation that Grossi and Velázquez-Quesada propose in their paper (similar to the one in van Benthem and Velázquez-Quesada 2010) seems appropriate for representing the awareness-of of human subjects.

They interpret the awareness-that also in a syntactic way and its semantics is given by the "acknowledgement function", that assigns a list of formulas to each agent at each world. In this case they state that the "acknowledgement of formulas does not need to have any closure property" (Grossi and Velázquez-Quesada 2015, p. 4077). Indeed, the list of formulas has no type of restrictions and the origin of the acknowledgement formulas stems from the application of some concrete epistemic action, such as a modus ponens step or an observation.

Awareness of and Awareness that: concepts used without theoretical foundation. As I hope has been made clear with the above analyses of two relevant papers, the concept of 'awareness' in its two variants has been widely employed, but poorly substantiated. This is one of the most clearest examples of the existing gap between works in contemporary Logics and the thoughts in Philosophy of Logic. As already mentioned, it is common in logical approaches to not make explicit the foundations of the employed concepts.

In many cases, this does not imply that the authors are not familiar with these foundations nor that they do not consider them to be important. It is rather a question of relevance and space limitations while preparing an academic paper. Since most of the important papers are directed to a specialized audience that, for the most part, has no philosophical interest, it is understandable that most authors do not include philosophical discussions in their contributions. The problem arises when this lack of foundation brings with it a clear devaluation of the theoretical questions, leading to a 'poor' understanding of many complex notions that are being used.

The lack of a coherent theoretical foundation may lead to misunderstandings and different interpretations of the same notions, thus diversifying their meanings and making different proposals about allegedly the same concepts, that are incomparable in their grounds. As showed here, this has been the case with 'awareness' but there are other concepts that suffer from the same 'problem'. The need of clarifying and defining in precise ways each and every complex notion that is being employed stands to reason because of the various applications these contemporary logics may have. Artificial Intelligence (AI) is a huge field of study, interdisciplinary in its roots, and that is being developed by experts from very distinct fields. The role of Logic in AI is clearly important and before applying it, I think that the foundations of its main concepts has to be unravelled.

To break the gap and tend bridges between the formal developments of contemporary logics and the philosophical considerations of their foundations seems like an advantageous task that will not only reconcile both fields of study, but also be of mutual benefit.

2.3 Combining Awareness-Of and Awareness-That

In the context of the aforementioned discussion it is time to take sides. I will thus proceed to define what I precisely mean with these expressions and advance my interpretation of the concept of 'knowledge'.

Awareness of. The information the agent has at her disposal, her perceptions, form the collection of her *awareness of*. This concept needs to be understood in the paradigm of dynamics of information, i.e. of information change. I understand awareness-of as a dynamic concept that will change constantly through the agent's communication process. An agent being aware of some information means that she is able to talk about this at the given moment. As such, one could understand it as the language she has at her disposal, or the 'working memory'.

Probably the most important feature of *awareness of* is that it can be formed by different types of information and the agent will not need to consider its truth or falsehood. For entertaining some given information she only needs to bear it in mind, but there is no restriction whatsoever on the kind of information it should be. As such, the collection of aware-of information can be formed, of course, by propositions, but also by doubts, mere perceptions, learned procedures or habits, etc.

This will be the weak point of my interpretation when trying to build a logical framework that captures the theoretical one, since though theoretically possible, it is not logically easy to formalize this. That is the reason why I will opt for a restriction to atomic propositions in the logical system, subscribing what (Grossi and Velázquez-Quesada 2015), among others, have proposed.

Awareness that. This notion is equated to the acknowledgement of the truth of some information. In some sense, it corresponds to a kind of explicit knowledge, even though I will draw a difference between both. Acknowledging the truth of a statement means having recognised it, but this recognition does not always need to be done *only* by the agent herself. As in the proposal from Grossi and Velázquez-Quesada (2015), the epistemic actions will play a fundamental role in the EAK-Schema, since they will determine the origin of the information the agent is aware-that.

Being aware-that some φ is the case is something that, though I will usually be considering the present moment, might have happened in the past. The whole collection of information the agent is aware-that will be present in the EAK-Schema, even if she is not entertaining it at the given moment. The epistemic action of *becoming aware of* will bring this information to the present moment.

Regarding the 'origin' of the aware-that information, since what the agent recognises is precisely the truth of this information, I will show two different ways (alluding to the epistemic actions the agent may perform) through which she will be able to reach it, namely, the actions of *deductive inference* and of *observation*.

Explicit knowledge as both awareness-of and awareness-that. The combination, as a conjunction, of the two types of awareness will give rise to my concept of 'explicit knowledge', which I will call 'Explicit Aware Knowledge' (EAK). Albeit the notion of 'awareness-that' has been (as in van Benthem and Velázquez-Quesada 2010) interpreted as 'explicit knowledge' in many occasions, I will argue that having recognised the truth of a given information at some moment is not enough. For 'really' constituting explicit knowledge, the information also needs to be entertained by the agent at the given point of evaluation.

This interpretation of 'explicit knowledge' will resemble the concept that Awareness Justification Internalism is aiming at. The combination of both awareness-of and awareness-that highlights that they are indeed different types of awareness and that together they can form a more complex understanding of the explicit information.

References

Ågotnes, T., & Alechina, N. (2014). A logic for reasoning about knowledge of unawareness. *Journal of Logic, Language and Information, 23*(2), 197–217.

Bergmann, M. (2006). *Justification Without Awareness: A Defense of Epistemic Externalism*. Oxford: Oxford University Press.

Dretske, F. (1993). Conscious experience. *Mind, 102*(406), 263–283.

Duc, H. N. (1997). Reasoning about rational, but not logically omniscient agents. *Journal of Logic and Computation, 7*(5), 633–648.

Fagin, R., Halpern, J., Moses, Y., & Vardi, M. Y. (1995). *Reasoning About Knowledge*. Cambridge, MA: The MIT Press.

Fagin, R., & Halpern, J. Y. (1988). Belief, awareness, and limited reasoning. *Artificial Intelligence, 34*(1), 39–76.

Fernández-Fernández, C. (2019). Informe bibliográfico sobre la lógica (epistémica) de la conciencia. *Contrastes. Revista de Filosofía, XXIV*(3), 105–120.

Gettier, E. (1963). Is justified true belief knowledge? *Analysis, 23*(6), 121–123.

Grossi, D., & Velázquez-Quesada, F. R. (2009). Twelve Angry Men: A Study on the Fine-Grain of Announcements. In X. He, J. F. Horty, & E. Pacuit (Eds.), *LORI-II* (Vol. 5834, pp. 147–160)., Lecture Notes in Computer Science Berlin, Heidelberg: Springer.

Grossi, D., & Velázquez-Quesada, F. R. (2015). Syntactic awareness in logical dynamics. *Synthese, 192*(12), 4071–4105.

Jago, M. (2009). Epistemic logic for rule-based agents. *Journal of Logic, Language and Information, 18*(1), 131–158.

Madison, B. J. C. (2009). On the compatibility of epistemic internalism and content externalism. *Acta Analytica, 24*(3), 173–183.

Quine, W. V. O. (1970). *Philosophy of Logic*. Cambridge, MA: Harvard University Press.

van Benthem, J. (2011). *Logical Dynamics of Information and Interaction*. New York: Cambridge University Press.

van Benthem, J., & Velázquez-Quesada, F. R. (2010). The dynamics of awareness. *Synthese, 177*(Supplement-1), 5–27.

van Ditmarsch, H., & French, T. (2009). Awareness and forgetting of facts and agents. In *Proceedings of the 2009 IEEE/WIC/ACM International Conference on Web Intelligence and International Conference on Intelligent Agent Technology—Workshops, Milan, Italy, 15–18 September 2009*, pp. 478–483. Washington, DC: IEEE Computer Society.

van Ditmarsch, H., & French, T. (2011). Becoming aware of propositional variables. In M. Banerjee, & A. Seth (Eds.), *Logic and Its Applications—4th Indian Conference, ICLA 2011, Delhi, India, January 5–11, 2011. Proceedings*, volume 6521 of *Lecture Notes in Computer Science*, pp. 204–218. Berlin: Springer.

van Ditmarsch, H., French, T., Velázquez-Quesada, F. R., & Wang, Y. N. (2013). Knowledge, awareness, and bisimulation. *TARK 2013. Proceedings of the 14th Conference on Theoretical Aspects of Rationality and Knowledge*, vol. 1, pp. 61–70.

van Ditmarsch, H., Halpern, J. Y., van der Hoek, W., & Kooi, B. (Eds.). (2015). *Handbook of Epistemic Logic*. London: College Publications.

Velázquez-Quesada, F. R. (2011). *Small steps in dynamics of information*. Ph.D. thesis, Institute for Logic, Language and Computation, Universiteit van Amsterdam.

Chapter 3
Knowledge as an Epistemological and Logical Concept

Abstract This chapter is devoted to the analysis of the concept of 'knowledge' both taken as an abstract notion and in its applications in Epistemic Logic. It starts with a review of the epistemological theories that have been developed since its inception. Afterwards it analyses the different interpretations that have been made of this concept in Epistemic Logic and its contemporary developments, highlighting the difference between 'implicit' and 'explicit' knowledge and the strategies for drawing this distinction. Finally, it offers a new understanding of these notions, which supports the correlation between human knowledge and 'Explicit Aware Knowledge' (EAK), the central notion of the theoretical framework: the EAK-Schema.

Keywords Explicit knowledge · Implicit knowledge · Epistemic internalism · Semantic approach · Explicit Aware Knowledge

3.1 Epistemological Definitions and Approaches to Knowledge

This section[1] is focused on an analysis of what 'knowledge' *really* could be and how it has been understood in the past decades. But before doing so, I will devote some pages to the definition of the concept itself.

Defining 'knowledge'. I will start this analysis with the 'standard' definitions given by two relevant dictionaries. As such, according to the *Oxford Dictionary* (OD), 'knowledge' means:

> 1. Facts, information, and skills acquired through experience or education; the theoretical or practical understanding of a subject. 1.1. The sum of what is known. 1.2. Information held on a computer system. 1.3. *Philosophy*. True, justified belief; certain understanding, as opposed to opinion.[2]

---

[1] For a general overview of Epistemology see the introductory Sect. 1.2.

[2] In https://www.lexico.com/definition/knowledge, accessed 02/20/2020.

C. Fernández-Fernández, *Awareness in Logic and Epistemology*,
Logic, Epistemology, and the Unity of Science 52,
https://doi.org/10.1007/978-3-030-69606-1_3

While according to the *Merriam-Webster Dictionary* (MWD), 'knowledge' is defined as follows:

> 1a(1): the fact or condition of knowing something with familiarity gained through experience or association. (2): acquaintance with or understanding of a science, art, or technique. b(1): the fact or condition of being aware of something. (2): the range of one's information or understanding. c: the circumstance or condition of apprehending truth or fact through reasoning: cognition. d: the fact or condition of having information or of being learned.[3]

Considering the different interpretations provided by these two definitions (respectively, only in its first meaning), it is easy to conclude that, in general, there is no 'consensus' about what 'knowledge' is or should be. In what follows, I will try to unravel what lies behind these interpretations.[4]

Given the first definition, (OD), 'knowledge' seems to be every information (understood here in a wide sense) that has been acquired. It alludes to the process of gaining knowledge, by saying "acquired through experience or education". This might create misunderstandings, since one can acquire through experience some types of information that are normally not equated with knowledge, but with beliefs.

In OD 'belief' is defined as "1. An acceptance that something exists or is true, especially one without proof. 1.1. Something one accepts as true or real; a firmly held opinion".[5] It is interesting to confront both these definitions, because the definition of 'knowledge' does not allude to the truth of the information (except for the philosophical meaning in 1.3), while the definition of 'belief' (in 1.1) does explicitly mention "one accepts as true". Being this so and attending to common sense intuitions, it seems that the only difference the OD establishes between knowledge and belief is based on the process of obtaining it, in the case of knowledge, "through experience or education" and in the case of belief through a personal acceptance.

Of course the definition of 'belief' also mentions "without a proof", alluding to the lack of 'justification' the believed information will have, but this strikes me still not enough to draw a clear difference between 'knowledge' and 'belief'. One can imagine having acquired true information by education, for example, and not being able to give a proof of the truth of it, will it then be knowledge or belief? I do not think that based on the definitions given by the OD there can be a convincing answer.

The meaning of 'knowledge' which refers to its philosophical provenance is also not so clear. Though it alludes to the JTB-view, it also employs the words "certain understanding, as opposed to opinion". One can convey, from a logical-philosophical point of view, that 'opinion' (*doxa*, in its Greek origin) can be the same as 'belief', drawing thus the definite division in the meaning of 'knowledge' and 'belief', but why should 'knowledge' always be opposed to 'opinion'? One can easily imagine cases where opinions or beliefs become knowledge, once the needed justification is achieved.

---

[3] In https://www.merriam-webster.com/dictionary/knowledge, accessed 02/20/2020.

[4] I refrain from considering other standard or specialised dictionaries, since I do not intend to make a lexicographic point. The use of these dictionary definitions, here and in other sections of this book, helps me relate my arguments with the common sense intuitions I wish to take into account.

[5] In https://www.lexico.com/definition/belief, accessed 02/20/2020.

To sum up, the definition given in OD does not clarify sufficiently the semantic content of 'knowledge', neither that of 'belief', with respect to what has been developed in Epistemology. Of course I assume that the definition given here by OD intends to capture a much wider semantic field, than that used in Epistemology. Nevertheless, it is interesting to notice how the *Oxford Dictionary* blurs the edges between the semantic field of 'knowledge' and that of 'belief'.

Regarding the definition given by the MWD, I want to mention that they start the definition of 'knowledge' with "the fact or condition of knowing". This means that here 'knowledge' is seen as a "fact or condition", while in OD it was defined (in 1) as "facts, information and skills". This seems to indicate that the authors from the OD consider a much wider semantic field than those from the MWD, since the former include "information" in their definition, while the latter do not. This point strikes me important, because what this chapter, among other goals, intends to do is to equate 'knowledge' with 'true information' and by leaving out exactly this concept, the MWD is contradicting my interpretation in the EAK-Schema.

What is also worthy to mention is that in the meaning 1b(1) they relate 'knowledge', i.e. 'the fact or condition of knowing', with "being aware of something". In my understanding, the mere fact of being aware of something refers to perceiving some information, irrespective of its truth or falsehood. But perceiving or entertaining information is not comparable to knowing it.

From this it follows that in comparison to the OD, the definition of 'knowledge' given in the MWD is narrower with respect to 'information', which is not mentioned explicitly; while also larger with respect to the 'awareness' of the subject, which is introduced as one possible meaning of knowledge.

Epistemology and Knowledge. I will now turn my attention to Epistemology in general. As explained in the introductory chapter, Epistemology can be considered the branch of Philosophy that studies knowledge. In the mentioned chapter, I reviewed the different epistemological views but I deliberately did not give a specific definition of knowledge. The various epistemological proposals result in a concrete understanding of knowledge, depending on what the subject does in order to gain this knowledge.

Nevertheless, there are several aspects of the concept of 'knowledge' that underlie and are presupposed by most epistemological views. The standard definitions of knowledge I analysed above show that the notion is polysemic and may refer to different ways of gaining or having true information. In Rescher (2003), the author introduces the concept by saying that "there are various significantly distinguishable sorts of knowledge in terms of the kind of thing that is at issue" (Rescher 2003, p. xiv).

He then (pp. xiv–xv) classifies four types of knowledge: (1) "knowledge-that", also called 'propositional knowledge', which is the standard understanding in Epistemology, referring to the knowledge of facts; (2) "Adverbial knowledge", meaning knowing what, when, how, etc.; (3) "Knowledge by acquaintance", alluding to 'individuals or things'; and (4) "performatory knowledge", concerning the knowledge of procedures or concrete actions.

With the above classification in mind it is now easier to accept the definitions given in the *Oxford* and the *Merriam-Webster Dictionaries*, since both tried to include, or at least take into account, all these types of knowledge. For example, when the definition 1b(1) in MWD uses the words 'aware of something', it is probably referring to knowledge by acquaintance, which would correspond to my interpretation as 'perceiving or entertaining some information'.

I agree with Rescher in the fact that Epistemology is mostly devoted to *propositional knowledge*, this being one of the most important assumptions that underlies almost all epistemological views and that is not always sufficiently clarified. In my case, the only type of knowledge suitable of fitting my purpose of creating a conceptual schema for tending bridges between Epistemology and an extension of Epistemic Logic is propositional knowledge, since it needs to be capable of a formalisation with atomic propositions. To the contrary, in other fields of study, like Philosophy of Science, the most prominent type of knowledge is adverbial knowledge (also called "tacit knowledge" or simply "knowledge-how"). Nowadays, the concept of knowledge-how is providing lively debates in this field.

Of course when it comes to defining in which sense the believed information is being justified for becoming knowledge, most epistemologists are considering propositional knowledge, that is, knowledge of facts that, in turn, need to have a truth value. But there are also other types of epistemological views, like those reviewed in Rescher (2003), where other forms of knowledge are taken into account. Though completely licit, it is not common for traditional Epistemology to consider them. As such, one might conclude (with the usual risk of drawing a general claim) that *Epistemology is traditionally concerned with propositional knowledge.*

Propositional knowledge ('knowledge of facts' or 'knowledge-that') is, according to Rescher, "coordinate with the capacity to answer questions, above all, in the case of knowledge-that-*p*, being in a position correctly and appropriately to answer the question: 'Is *p* true or not?'" (Rescher 2003, p. xv). Being this so, every epistemological view or perspective is devoted to the answer of whether the known object 'is true or not'. Upon that, an epistemological approach considers the possibility of the truth and provides a structure to the various origins from which the subject comes to this truth.[6]

Knowledge in Awareness Justification Internalism. To conclude this section I shall devote some words to the meaning of 'knowledge' in Awareness Justification Internalism (AJI), the view I chose in Sect. 1.2 and upon which I will construe my theoretical framework. In line with what I said above, AJI is of course also concerned with propositional knowledge, but in different forms, and that is the key point of this view.

In AJI, propositional knowledge is the result of the process through which the agent transforms her true belief into knowledge with the 'help' of her awareness of other information, that acts as a justifier. In other words, if the agent has a belief about something, i.e. if she entertains some information whose truth she has not yet acknowledged, and if she is aware of some other information and can come to the

[6]A review of these views and a personal choice is provided in Sect. 1.2.

truth of her belief through this aware information, then, once she is aware that this information is indeed the case, she will know it.

In the above definition I tried to summarise the AJI-view mentioning all its components, let me now unravel them one after another:

- *Belief:* for the purpose of this research, belief will be all information the agent entertains and is eligible for becoming knowledge by the application of the most convenient epistemic action. As such, a believed information is equated with information the agent is *aware of.*
- *Truth:* the acknowledgement of the truth of a given information turns this information automatically into *awareness-that*. For coming to this truth there will be two possible epistemic actions, those of 'deductive inference' and 'observation'. The truth of the information needs to be reached by the agent through an epistemic action.
- *Knowledge:* true information the agent is aware-that it is the case and can justify by the recognition of the epistemic actions she has performed and that made the aware-of information become aware-that.
- *Justification:* the notion of 'justification' lies in the capability of the agent of performing epistemic actions and being able to recognise and name them.

Taking this view as an example of any other epistemological approach, notice that it does not really consider *what* knowledge is, but rather *how* knowledge is obtained by an agent. The definition of what makes something be 'knowledge' is of course influenced by the answer of how it becomes what it is, but it is surely not the same. Thus, I shall distinguish the epistemological definitions of knowledge (the 'how') from the theoretical or philosophical definitions (the 'what').

In what follows and regarding Epistemology, I will assume the AJI view as explained here and presuppose, as most epistemologists do, that the knowledge at stake is always 'propositional', that is, it is expressible in logically meaningful terms.

3.2 Explicit Knowledge and Implicit Knowledge: A Key Distinction

I have established that the type of knowledge that concerns Epistemology is *propositional knowledge*. Regarding Epistemic Logic, I shall state the same, since what is considered is always a form of 'knowing-that-φ is true'. And, of course, there are different logical interpretations of how this 'knowing-that' might be modelled, represented and formalized.

This section will review some possible representations of knowledge in Logic, starting with (Hintikka 1962), explaining the distinction between 'implicit' and 'explicit knowledge' (or information) and presenting the different strategies for obtaining this distinction.

Knowledge in Logic: Standard Epistemic Logic. In the formal structure that Hintikka (1962) develops, presented in Sect. 1.1, the author employs the semantic of possible worlds and takes the structure of modal logic as the foundation of Epistemic Logic, with the knowledge operator being analogous to the modal necessity operator.

Regarding the sense in which he defines knowledge, one could say, that without explicitly mentioning it, he is describing 'propositional knowledge' in the *Prolegomena* to his work. He says that his intentions are to "defend criteria of consistency for certain sets of statements [. . .] they are statements made in terms of the following forms of words: (1) '*a* knows that *p*.' (2) '*a* knows whether *p*.' (3) '*a* does not know that *p*'. [. . .]" (Hintikka 1962, p. 3).

It is convenient to recall, that Hintikka provides a structure for both Epistemic Logic and Doxastic Logic, i.e. logic of knowledge and logic of belief, respectively. As such, with (1) he refers to statements about the knowledge of the agents and with (2) to statements about their belief. In his *Prolegomena* he introduces the main concepts he will employ throughout his work, but devotes little space to a proper theoretical definition of 'knowledge' itself.

He draws an important distinction between "statements" and "sentences", stating that "a statement is the act of uttering, writing, or otherwise expressing a declarative sentence. A sentence is the form of words which is uttered or written when a statement is made" (Hintikka 1962, p. 6). These "declarative sentences" are the different forms he presents on page 3 (see previous quote) and which are "said to express epistemic notions" (Hintikka 1962, p. 4).

He continues his *Prolegomena* limiting the form of the statements he will be concerned with and establishing some initial criteria of consistency. In general, he establishes the monotonicity of his logic, that is, that creating new knowledge can never change what is already known. This feature makes Epistemic Logic completely static and unable to capture changes in knowledge (like subsequent logics will do).

Epistemic Logic represents, in a nutshell, the invariant propositional knowledge that can be expressed with declarative statements. Not only is it invariant, but it is also 'ideal' or 'omniscient' as already addressed in Sect. 1.1. There, I mentioned that one of the possible solutions to prevent the agents from being logically omniscient consists in splitting knowledge into 'explicit' and 'implicit'.

Distinction: *Implicit* vs *Explicit* Knowledge. So far, the distinction between 'implicit' and 'explicit' knowledge should be clear, but I will devote some more words to it before showing the different strategies that can be followed to obtain it.

Implicit Knowledge is considered the ideal knowledge an agent would obtain if she were able to perform all possible inferences with what she knows. In other terms, one could say that implicit knowledge is the best an agent can do if she were omniscient. It is a very useful concept regarding Artificial Intelligence and different computational applications of Epistemic Logic, but it must be considered a 'logical construct' from a purely theoretical point of view. Whilst *Explicit Knowledge* represents what the agent really knows at a given moment. Explicitly knowing corresponds, in this way, to the intuitive understanding of a human subject knowing some information.

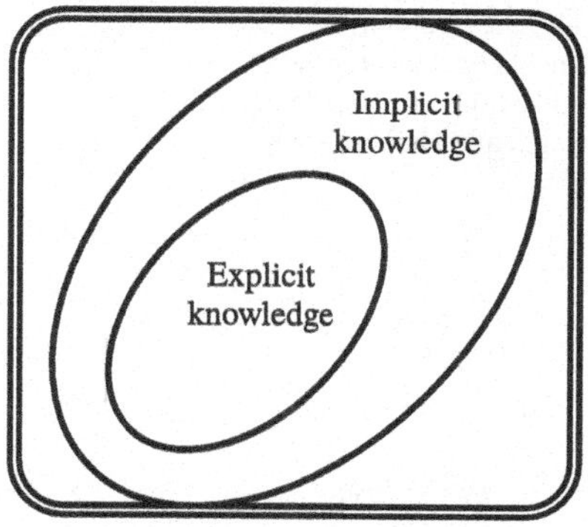

Fig. 3.1 Implicit and explicit knowledge in Konolige (1984)

There are two main strategies that can be followed for drawing the distinction: (*a*) the *syntactic* approach and (*b*) the *semantic* approach, which usually incorporates some external concept. These two types depend on the structure one decides to apply on the relation between the information and the agent, which can be seen from a (*a*) sentential or a (*b*) propositional point of view.

The **syntactic approach** can be represented by a *deductive system* as the one proposed in Konolige (1984), based on the ideas from Levesque (1984). It is a purely syntactic system, since the information (here the author considers 'belief' instead of 'knowledge') is represented as a list of sentences that is ascribed to an agent. In contrast to the possible worlds approach, here the possibility of creating new information is given by a set of procedures (inference rules) that is combined with the set of information the agent possesses. This approach has its roots in the 'rule-based systems' from AI and, according to the author, is "compatible with current philosophical theories of human cognitive states" (Konolige 1984, p. 8).

The deduction model is applied in Konolige (1986) for representing both implicit and explicit 'belief' (notion that I will take her as refering to 'information' in a broad sense). In this sentential view explicit knowledge is defined directly as a set of sentences, i.e. as the primitive knowledge the agent has. Implicit knowledge is then what follows (deductively) from what is explicitly known.

In Fig. 3.1, a diagram taken from Konolige (1986), the small ellipse represents the explicit or real knowledge of the agent, defined as a list of sentences; while the bigger ellipse, that includes the small one, represents everything that can be deduced from the explicit knowledge. The rectangle represents all true information.

In words more familiar to the dynamics of information one could say that such interpretation assumes that the only action available to the agent is *deductive inference*. It is precisely the possibility of performing this inference what creates the implicit knowledge set. The main feature in this approach is that the concept of *explicit knowledge* is prior in the system and *implicit knowledge* is created from it. As the author states it: "the explicit beliefs are normally considered to be primary, that is, they represent the beliefs of the agent, and the implicit beliefs are derived from them by closure under logical consequence" (Konolige 1986, p. 243).

Fig. 3.2 Implicit knowledge, awareness-of and explicit knowledge in Fagin and Halpern (1988)

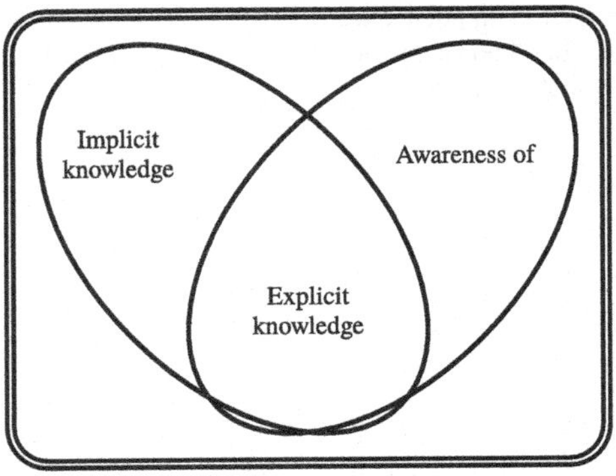

The **semantic approach** bases the relation between the agent and the information she entertains on the ways of obtaining the truth of this information. One of these approaches, which I will take as paradigmatic, is the already mentioned *Awareness Logic* (AL) presented in Fagin and Halpern (1988), that extends the structure of EL. *Explicit knowledge* is defined here as the implicit knowledge the agent is aware of.

The authors take the implicit knowledge from standard EL as their starting point (or primary notion). By adding the awareness of the agent as a syntactic operator (a list of formulas defined as the information the agent is aware of), which acts as a filter on the implicit knowledge, they obtain explicit knowledge. In this sense, explicit knowledge is the result of the combination of two other concepts, namely, implicit knowledge and awareness-of.

It is worthwhile noticing that what is here called "implicit knowledge" is not what follows from the explicit knowledge of the agent, as it was in the previous approach, but rather what the agent would know explicitly if she were aware of every formula that is true in all her epistemic possibilities, i.e. if she was logically omniscient.

Figure 3.2 (also taken from Konolige 1986), shows the relation between the notions of 'implicit knowledge', 'awareness of' and 'explicit knowledge' in AL by Fagin and Halpern (1988). As in the previous figure, the rectangle represents all true information. The ellipse on the left pictures implicit knowledge, while the ellipse on the right stands for the awareness-of of the agent. The intersection of both these sets gives rise to explicit knowledge.

Regarding the dynamics of information, in this approach the only possibility of changing the given information is by changing the 'size' of the awareness set and this is done by an action like *becoming aware* or *becoming unaware*. The possibility of an action of deductive inference, as in the previous case, is not given, since the explicit knowledge is defined in terms of the two other concepts.

The critique that Konolige carries out in his paper can be summarised with the following quote: "[t]he logic of general awareness thus characterizes agents as perfect reasoners, restricted in some way to considering a subset of possible sentences to reason about" (Konolige 1986, p. 245). In other words, he is describing that the agents from AL are still omniscient in some sense ("perfect reasoners") and their capability of having 'real knowledge' is ascribed to them through the list of what constitutes their awareness.[7] I do partially agree with this critique, in the sense that I do not share the choice Fagin and Halpern make in taking EL as it is. Of course, AL has been very inspiring, but the essence of logical omniscience is still represented here.

In conclusion, these two approaches for splitting knowledge into 'implicit' and 'explicit' provide two distinct views on the relation between the agent and the information ascribed to her. From a theoretical point of view, the main difference lies in the choice of which concept is taken as the *primary notion* in the system. From this perspective, the sentential approach is probably more coherent with the common sense intuitions, since it takes what the agent really knows as its primary notion. While the semantic approach is not that intuitive regarding the primary concept, it does include the awareness of the agent in its system, which is a point in its favour. Of course, AL is not the only possible semantic approach for distinguishing between implicit and explicit knowledge as will be showed in what follows.

Another relevant proposal that distinguishes between implicit and explicit information, but uses a different strategy is Epistemic **Justification Logic** (JL) (Artemov 2008; Renne 2012; Artemov and Fitting 2016), where explicit knowledge is defined as the implicit knowledge the agent has a justification for. The strategy could be seen as analogous to that from Awareness Logic, since it incorporates an external concept, namely 'justification', through which the explicit information is elaborated, but the mechanisms for establishing the distinction are quite different.

JL stems from the idea of bringing the 'justification' from 'Justified True Belief' (JTB) into Epistemic Logic (EL) for reconciling EL with Epistemology. This logic is inspired by the Logic of Proof (LP) (which can be traced back to the early 1930s intuitionism in mathematics and its proof-based semantics like, for example, in Kolmogorov 1932 and Heyting 1934). The keystone of JL is that it adds 'justification terms' (variables or constants) to the language of EL, such that the resulting formulas that include these 'justification terms' correspond to explicitly justified formulas. As such, '$t : F$' means that *t is a justification for formula F* and the resulting knowledge is then a 'justified explicit knowledge'.

For the purpose of this chapter, it should suffice to highlight that JL is a very interesting proposal that also distinguishes between implicit and explicit information by adding the 'justification' of a formula to the logical structure of EL. The standard system from EL is maintained and remains 'implicit' and therefore omniscient, while the 'new' structure from LP is introduced for creating the explicit counterpart. An information will be explicitly known whenever the agent has a justification for it that can be shown.

The mathematical structure of LP is quite more complicated than the standard modal semantics from EL, thus, the combination of both provides a very coherent and embracing approach. Though JL can also be combined with the concept of

[7] Note that Konolige's critique might be applied to his own proposal, since the agents he describes reason on the basis of an incomplete set of inference rules (procedures) and, thus, are also "perfect reasoners" in some sense.

'awareness' (see Baltag et al. 2014 or Sedlár 2013, for example), giving rise to even more complex structures with some interesting properties, I will not take these type of solutions into account regarding the EAK-Schema. The branch of Justification Logics is probably still too much computationally oriented and I think that their agents are still closer to perfect reasoners with some limitations, than to human subjects.

Knowledge in Dynamic Epistemic Logic. In Sect. 2.2 I reviewed the different uses and applications of the concept of 'awareness' in two recent papers. While doing so, I mentioned how the resulting concepts of 'implicit' or 'explicit' knowledge were represented. I will now pick up on that in order to clarify in which sense these two papers structure the concept of 'knowledge' and afterwards review a third paper that does not incorporate the awareness of the agent, but follows a quite different strategy for splitting knowledge into its implicit and explicit counterpart.

What the following three papers have in common is that they can all be classified as different approaches in the field of *Dynamic Epistemic Logics*[8] and as such, the role of the epistemic actions is crucial. Though they constitute a static representation of knowledge at each moment, with the appropriate epistemic actions, the system is able to change some convenient part of the model to obtain an updated version of it and represent the information change. The actions are what prevent the agents from being omniscient, in correspondence to the definitions of knowledge.

(1) In "The dynamics of awareness", (van Benthem and Velázquez-Quesada 2010), the authors take AL from Fagin and Halpern (1988) as their basis and include some concrete modifications. The most relevant modification concerns explicit knowledge (Ex), which is now defined as the implicit knowledge (K) of both a formula (φ) and the awareness of this formula ($A\varphi$), that is $Ex\varphi := K(\varphi \wedge A\varphi)$.

The implicit knowledge is still the standard from EL, while explicit knowledge is equated with 'awareness that'. Once the authors present their framework (epistemic awareness models) they engage into the topic of the paper, i.e. "The dynamics of awareness". For doing so, they state that "[t]hough the agent is not logically omniscient anymore, she can get new information by various acts, including observation and inference. [...] it also makes sense to look for simple actions transforming models that can be put together to analyze more complex informational acts" (van Benthem and Velázquez-Quesada 2010, p. 10).

In what follows, van Benthem and Velázquez-Quesada present the operations of *consider*, *drop*, *implicit observation* and *explicit seeing*. The first two operations refer to the agents awareness-of set, being it extended in the case of 'consider' and shrunk in the case of 'drop'. The next action changes the whole model, affecting thus the implicit knowledge of the agent. After an implicit observation the agent will discard all "those worlds where some *observed* formula χ fails" (van Benthem and Velázquez-Quesada 2010, p. 10). The last one, 'explicit seeing', is a combination of the implicit observation and an act of consideration.

[8]They also correspond to authors from the same tradition. I could have chosen other papers for this review, but since part of my own research has been in collaboration with authors from this tradition, I preferred to take their work as paradigmatic examples of the concepts I am analysing.

The crucial role of the epistemic actions comes from the fact that they show how the informational attitudes of the agent may change and what part of the model (meaning, what set of her information or relation) will be affected. The actions of 'consider' and 'drop' only affect the awareness set, this implies that the agent has the possibility of entertaining new information or 'forgetting' some of the information she was considering. The 'implicit observation' is, of course, an idealized action, but logically possible. What makes the agents definitely not omniscient (though still almost perfect) in this approach is the definition of explicit knowledge and the possible effects the presented actions might have on it. Thus, for an agent to extend her explicit knowledge (explicit seeing operation), she will need to consider some new information and also perform an implicit observation of it.

(2) In "Syntactic awareness in logical dynamics" (Grossi and Velázquez-Quesada 2015) the framework is more complex than in the previous approach. The authors present a language, based on EL, that also includes 'availability' and 'acknowledgement' formulas (as reviewed in Sect. 2.2). The semantic model they consider is formed by the standard semantic model from EL (with the corresponding subscripts for each agent in their multi-agent model) and two more functions: (i) a 'propositional availability function', $\mathtt{PA}_i$, "returning the set of atomic propositions agent $i \in \mathtt{Ag}$ has available at each possible world"; and (ii) an 'acknowledgement set function', $\mathtt{AC}_i$, "returning the set of formulas agent $i \in \mathtt{Ag}$ has acknowledged as true at each possible world" (Grossi and Velázquez-Quesada 2015, p. 4077).

These functions create the awareness-of (availability) formulas, ${}^{av(i)}\varphi$, and the awareness-that (acknowledgement) formulas, ${}^{ack(i)}\varphi$, respectively. The main difference regarding their semantic interpretation is that the availability formulas are part of a set of atomic propositions, while the acknowledgement formulas belong to a set of formulas, i.e. "the former asks for φ's atoms to be available to the agent, the second ask[s] for φ itself to be acknowledged" (Grossi and Velázquez-Quesada 2015, p. 4077).

Starting from this framework, the authors present the main 'epistemic notions' they consider and their definitions, i.e. the conditions that the agent needs to fulfil in order to obtain the information in the defined sense. In Table 2 (p. 4079) they define the two types of awareness: *awareness of* ($Ao_i(\varphi) := \Box_i^{av(i)} \varphi$) and *awareness that* ($At_i(\varphi) := \Box_i^{ack(i)} \varphi$); and three types of knowledge:

(i) Implicit knowledge: $Imp_i(\varphi) := \Box_i({}^{av(i)}\varphi \wedge \varphi)$,
(ii) Inherent knowledge: $Inh_i(\varphi) := \Box_i({}^{ack(i)}\varphi \wedge \varphi)$ and
(iii) Explicit knowledge: $Exp_i(\varphi) := \Box_i({}^{av(i)}\varphi \wedge \varphi \wedge^{ack(i)} \varphi)$

Notice that in the three definitions the implicit knowledge operator (the box) is always outside the parenthesis, being thus applied to everything that follows. As such, with (i) *implicit knowledge* the authors refer to "what is true in all the worlds the agent considers possible and can be expressed in her current 'working language'" (Grossi and Velázquez-Quesada 2015, p. 4079). What is different to the same concept in previous approaches is that now the agent needs to be aware-of, i.e. needs to be considering or entertaining at this moment, the information she implicitly knows.

Since it is implicit, it is closed under logical consequence, as the authors show, and corresponds thus to an idealized knowledge.

With respect to (ii) *inherent knowledge*, the new notion they introduce in comparison to the previous views I have reviewed here, they state that it corresponds to "what is true in all the worlds the agent considers possible and has been accepted in her 'working memory'" (Grossi and Velázquez-Quesada 2015, p. 4080). This requirement refers to the awareness-that of the agent, i.e. she needs to have acknowledged the truth of the formula at some point, irrespective of her current entertainment of it or not.

Regarding (iii) *explicit knowledge* they propose to consider it as the union of both 'implicit' and 'inherent' knowledge. They state that "[f]or φ to be explicitly known, all worlds the agent considers possible must satisfy φ, and the agent must be aware *that* and *aware of* φ" (Grossi and Velázquez-Quesada 2015, p. 4080. Original italics).

Though the background idea of combining both awareness-of and awareness-that for obtaining 'explicit knowledge' strikes me the most successful one, as will be made clear at the end of this chapter, I cannot accept two features of their proposal: (a) they leave the standard knowledge operator ($\Box_i$) ranging over the scope of every other item in the definition (outside the parenthesis). This implies that for every type of knowledge, the formula is always true in every possible world the agent considers, and this is a very strong assumption. This feature also shows that the main structure from EL is still untouched and that they built their system upon it.

(b) The second feature I do not agree with from their proposal is the fact that they take awareness-of and awareness-that as two different sets formed in a similar way (syntactic consideration), that is, they consider both types of awareness as 'subsets' of the 'awareness set' from the original proposal in Fagin and Halpern (1988). They say that "[i]n the present analysis, the awareness set of awareness logic is split in two: the availability set, modelling the 'working language' of the agent [...] and the acknowledgement set, modelling what the agent has recognised as true" (Grossi and Velázquez-Quesada 2015, p. 4089). While this resource is useful for their purposes, it strikes me counterintuitive from a theoretical perspective. With awareness-of being a way of perceiving information and awareness-that being the acknowledgement of the truth of some other information, I do not think that a list of formulas can represent the latter as it is supposed to. The authors relate awareness-that with the concept of 'working memory', but I think that it would be more appropriate to equate it with the notion of 'knowledge memory', since in my understanding of the concept, awareness-that is already a type of knowledge.

Regarding the dynamics of this approach, the agents define three epistemic actions: 'becoming aware', 'deductive inference' and 'announcement'. The actions they describe will prevent the agents from logical omniscience, as in the previous approach, by modifying some (or all) elements of the defined model. As such, the action of *becoming aware* will modify the 'propositional availability function' and extend thus the set of 'availability formulas'. The action of *deductive inference* will modify the 'acknowledgement set function', adding to the set of 'acknowledgement formulas' those that follow from a *modus ponens inference step*. Lastly, the action of *announcement* modifies the whole model, restricting the accessibility relation to those worlds where the announced formula is true (similar to what is done in Public Announcement Logic, Plaza 1989, 2007).

(3) In "Explicit and Implicit Knowledge in Neighbourhood Models", (Velázquez-Quesada 2013), the author follows a different strategy for distinguishing between implicit and explicit information, based on a different semantic structure. So far, every approach I have reviewed (except for the purely syntactic approach) has maintained, in some sense, the original semantic structure from Fagin and Halpern (1988), but there have been other proposals that changed this structure, inspired by model-theoretic techniques. Though some of these approaches, like (Modica and Rustichini 1999) and (Heifetz et al. 2006), have been shown to be reducible to the original approach in AL, it does not mean that this line of research should be abandoned.

Therefore, in Velázquez-Quesada (2013) the author takes over this tradition of employing neighbourhood semantics and applies it to the distinction between implicit and explicit knowledge. As advanced before, this approach does not include the notion of 'awareness' (though it would be capable of, as shows the formal model I will propose), but is nevertheless interesting, since it does include the background intuition I highlighted from the syntactic approach, namely, that explicit knowledge should be considered the 'primary notion' in the system and also adopts a dynamic perspective on information. I will review the basics from this approach, trying to avoid a huge degree of technicality.

Neighbourhood models as deployed in Pacuit (2017) or Chellas (1980) (see Chap. 7 for a detailed presentation) are semantic structures that develop the traditional relational structure (Kripke semantics). Instead of building a model based on a set of worlds, W, an accessibility relation among them, R, and a valuation function, V; in a neighbourhood structure "[e]ach state from W is associated with a subset space over W" (Pacuit 2017, p. 6). The relation that is established is the *neighbourhood function*, N, which assigns a *set of sets of worlds* to each world $w \in W$, $N : W \to \wp(\wp(W))$ and needs to fulfil the *truth property*: "for every $w \in W$, if $U \in N(w)$ then $w \in U$" (Velázquez-Quesada 2013, p. 241). This property assures that for each subset of the neighbourhood of a given world, this world also belongs to it. The valuation function V, assigns then the true propositions to a set of worlds, $V : \mathrm{P} \to \wp(W)$. With these components, a *knowledge neighbourhood model* is a tuple $M = \langle W, N, V\rangle$ where $W \neq \varnothing$, N is the neighbourhood function and V the valuation function.

Regarding the semantic interpretation, a formula $\varphi \in \mathcal{L}$ is evaluated with its *truth set*, $[\![\varphi]\!]^M$, that contains the worlds in M where φ is true. The recursive definitions for the standard boolean operators are relations between these sets (for example, the intersections of truth-sets for the conjunction). The relevant semantic interpretation in this approach is the clause for the knowledge operator: $[\![\Box\varphi]\!]^M := \left\{w \in W \mid [\![\varphi]\!]^M \in N(W)\right\}$, which says that "a world w is in $[\![\Box\varphi]\!]^M$ (i.e. $\Box\varphi$ is true at w) if and only if the set of worlds where φ is true, $[\![\varphi]\!]^M$, is in the neighbourhood of w, $N(w)$" (Velázquez-Quesada 2013, p. 241).

The main reason that the author mentions for employing these type of structures is that the knowledge operator, $\Box$, is understood here as *explicit knowledge*. The reasons for this fact depend on the neighbourhood structure itself. The knowledge depicted here does not contain every validity, nor is it closed under modus ponens (as it was in relational models), the only closure property it has is *closure under logical equivalence*, meaning that "if φ and ψ are true in the same worlds of every model [...], then the agent knows the former if and only if she knows the latter" (Velázquez-Quesada 2013, p. 242). The author points out that, although this closure property suggest an idealised agent, it is always less idealised than the logical omniscient agents. Furthermore, if this approach is extended with further components (such as awareness or justification terms) this property can be avoided.

So far, Velázquez-Quesada has provided a semantic model that has *explicit knowledge* as its primary notion. For obtaining implicit knowledge he will follow the background intuition from the standard syntactic approaches, that is, "implicit knowledge is defined as the closure under logical consequence of its explicit counterpart" (Velázquez-Quesada 2013, p. 243). For obtaining this closure property he applies a known technique to 'convert' neighbourhood models into logically omniscient relational models (standard in EL). Since the neighbourhood structure is just an 'extension' of the relational structure, there is a 'way back', i.e. "a relational model can be seen as a neighbourhood model in which the neighbourhood function satisfies certain closure properties" (Velázquez-Quesada 2013, p. 243).

Thus, the N needs to fulfil these properties, which I will not describe in detail: closure under supersets, closure under binary intersections and containing the unit (for a deeper description of these properties and their proofs see Chellas 1980 and Pacuit 2017). Once a finite neighbourhood model satisfies these properties, it behaves as a relational model. For being able to apply this 'change' from neighbourhood to relational semantics, the author establishes a *full closure modality*, $[*]$, that behaves similar to any epistemic action, in the sense that it changes parts of the original model to obtain a resulting model where the information has been altered (in particular, the neighbourhood function has changed, including now the mentioned closure properties).

Using this 'closure modality' he is now able to describe both explicit and implicit knowledge: explicit knowledge will be the standard knowledge defined in neighbourhood models, $\mathrm{K}_{Ex}\,\varphi := \Box\,\varphi$, and implicit knowledge will be the closure of the former, $\mathrm{K}_{Im}\,\varphi := [*]\,\Box\,\varphi$. This approach strikes me not only important and interesting, but also very intuitive regarding its background intuitions, namely, that explicit knowledge is the primary notion that needs to be formalised (the real knowledge of 'real' agents) and implicit knowledge is what 'follows' from it or what an agent with unlimited reasoning abilities would obtain.

The author includes one epistemic action, "a Modus Ponens action" that is interpreted as a deductive inference step and serves to extend what the agent knows explicitly. Since the standard knowledge operator ($\Box$) is interpreted here as explicit knowledge, it is possible that an agent knows explicitly $\varphi \rightarrow \psi$ and φ, but does not know ψ explicitly, since this knowledge is not closed under logical consequence. The Modus Ponens action formalises the step to obtain ψ in this case. As such, if the Modus Ponens modality, $[{}^{\eta}\!\hookrightarrow_{\chi}]$, is applied, it will change the neighbourhood function in the resulting model. The new neighbourhood function will include the truth set of the consequent formula of the implication, whenever it is guaranteed that the truth sets of the implication itself and the antecedent are already part of the neighbourhood before the application of the action.

The agents presented in this approach are non-omniscient, regarding their explicit knowledge and the action of Modus Ponens provides a way to capture single deductive inference steps that the agent may perform. It is also worth highlighting, that this approach can be extended with several other actions, such as an action of 'becoming aware'.

3.3 Explicit Aware Knowledge (EAK) as the Core Concept

In the previous section I reviewed some proposals that have split knowledge into 'implicit' and 'explicit' following different strategies and I have described some of them as more intuitive than others. Based on that, my interpretation of these notions for the EAK-Schema will be the following:

- *Explicit Knowledge:* regarding real agents, that is, human beings and computing machines, who have limited reasoning abilities (because of space and capacity reasons) the immediate knowledge that can be ascribed to them is, in some sense, needed of limitation. An idealised notion does not fit with my main goal of tending bridges between epistemological and logical considerations. Departing from the knowledge that Awareness Justification Internalism (AJI) presents, its logical 'counterpart' needs to include the concepts of true information, awareness and justification. Furthermore, a notion of *awareness of* needs to be guaranteed, in the sense that it has to reflect what the agent is considering at a given moment, as well as a notion of *awareness that*, meaning that the agent needs to have acknowledged the truth of the information. It will also need to include a notion of *justification* which will be addressed in the next chapter.
  All the above leads me to define **Explicit Aware Knowledge** as the *information an agent is considering at a given moment, she is aware-of, and whose truth she has acknowledged, she is aware-that.* These two features are not interrelated, in the sense that she might be aware-of more information than the one she is also aware-that and that she might have been aware-that some information is true, but not be considering it at a given moment or point of evaluation.
- *Implicit Knowledge:* with this concept I will describe 'the best an agent can do', i.e. what she would obtain if she were a perfect reasoner. Though AJI does not consider this concept as such, it does mention the expression "potentially aware" in the definition of the Awareness Requirement (see Sect. 1.2). A 'potential' known information is what I will consider implicit knowledge, since it is logically defined as the closure under logical consequence of explicit knowledge, provided that what the agent will know implicitly are all the logical consequences of her explicit knowledge Specifically, I will consider the closure of the *awareness that* of the agent, that is, everything she might deductively infer from what she is aware-that

(including the aware-that information itself) at a given moment. This concept is also theoretically useful, since it describes the possible information an agent could obtain departing from what she explicitly knows and helps creating a bigger picture of all the information that is at her immediate scope.

As should be clear, the EAK-Schema is inspired, first, by the background intuition from the syntactic approach in Konolige (1984), i.e. explicit information is the primary notion in the system and implicit information is what follows from it; and second, by the brilliant idea from Fagin and Halpern (1988) on incorporating the concept of 'awareness', though I will apply two distinct types of awareness, namely, awareness-of and awareness-that.

The approach in Grossi and Velázquez-Quesada (2015) follows a quite similar strategy in the definition of the explicit knowledge as a combination of both types of awareness, but, as pointed out above, they do not understand 'awareness-that' in a semantic sense (from a knowledge-like perspective) as I do, neither do they take explicit information as the primary notion in their system. Their proposal is, regarding 'awareness', purely syntactical. They define both 'awareness-of' and 'awareness-that' as a list of information, atomic propositions in the former case and formulas in the latter. While I agree with the syntactic interpretation of 'awareness-of', since it represents intuitively the 'language' the agent has at her disposal, I do not agree with the representation of 'awareness-that'. I defined previously 'awareness-that' as that information whose truth the agent has acknowledged, this acknowledgement implies a semantic understanding of the concept, more related to 'knowledge' than to 'awareness-of'.

I do not imply hereby that their system does not solve the problem of logical omniscience, nor that it is 'worse' or 'less apt' than my proposal. What I am intending to highlight is that from a purely theoretical point of view and having in mind the bridge I want to build between these concepts and the epistemological view (AJI), the treatment of 'awareness-that' needs to be semantic, that is, the formal interpretation needs to show the truth of the information and the ways of gaining it. A semantic interpretation, in comparison to a syntactical one, will better reveal how the relation between the agent and the truth of her information is constructed. This being so, I will diverge from the proposal in Grossi and Velázquez-Quesada (2015) in their understanding of 'awareness-that', while maintaining the syntactic interpretation of 'awareness-of' as well as the interpretation of explicit knowledge as *awareness-of* plus *awareness-that*.

References

Artemov, S. (2008). The logic of justification. *The Review of Symbolic Logic*, *1*(4), 477–513.

Artemov, S., & Fitting, M. (2016). Justification logic. In E. N. Zalta (Ed.), *The Stanford Encyclopedia of Philosophy*. Metaphysics Research Lab, Stanford University, winter 2016 edition.

Baltag, A., Renne, B., & Smets, S. (2014). The logic of justified belief, explicit knowledge, and conclusive evidence. *Annals of Pure and Applied Logic*, *165*(1), 49–81.

Chellas, B. F. (1980). *Modal Logic: An Introduction*. Cambridge: Cambridge University Press.
Fagin, R., & Halpern, J. Y. (1988). Belief, awareness, and limited reasoning. *Artificial Intelligence*, *34*(1), 39–76.
Grossi, D., & Velázquez-Quesada, F. R. (2015). Syntactic awareness in logical dynamics. *Synthese*, *192*(12), 4071–4105.
Heifetz, A., Meier, M., & Schipper, B. C. (2006). Interactive unawareness. *Journal of Economic Theory*, *130*(1), 78–94.
Heyting, A. (1934). *Mathematische Grundlagenforschung. Intuitionismus. Beweistheorie*. Berlin: Springer.
Hintikka, J. (1962). *Knowledge and Belief*. Ithaca: Cornell U.P.
Kolmogorov, A. (1932). Zur Deutung der intuitionistischen Logik. *Mathematische Zeitschrift*, *35*, 58–65.
Konolige, K. (1984). *A Deduction Model of Belief and its Logics*. Ph.D. thesis, Computer Science Department, Stanford University.
Konolige, K. (1986). What awareness isn't: A sentential view of implicit and explicit belief. In *Proceedings of the 1st Conference on Theoretical Aspects of Reasoning about Knowledge, Monterey, CA, March 1986* (pp. 241–250).
Levesque, H. J. (1984). A logic of implicit and explicit belief. In R. J. Brachman (Ed.), *Proceedings of the National Conference on Artificial Intelligence. Austin, TX, 6–10 Aug. 1984* (pp. 198–202).
Modica, S., & Rustichini, A. (1999). Unawareness and partitional information structures. *Games and Economic Behavior*, *27*(2), 265–298.
Pacuit, E. (2017). *Neighborhood Semantics for Modal Logic*. Berlin: Springer.
Plaza, J. (2007). Logics of public communications. *Synthese*, *158*(2), 165–179.
Plaza, J. A. (1989). Logics of public communications. In M. L. Emrich, M. S. Pfeifer, M. Hadzikadic, & Z. W. Ras (Eds.), *Proceedings of the 4th International Symposium on Methodologies for Intelligent Systems* (pp. 201–216). Tennessee: ORNL.
Renne, B. (2012). Multi-agent justification logic: Communication and evidence elimination. *Synthese*, *185*(Suppl. 1), 43–82.
Rescher, N. (2003). *Epistemology: An Introduction to the Theory of Knowledge*. New York: State University of New York Press.
Sedlár, I. (2013). Justifications, awareness and epistemic dynamics. In S. N. Artëmov, & A. Nerode, (Eds.), *Logical Foundations of Computer Science, International Symposium, LFCS 2013, San Diego, CA, USA, January 6–8, 2013. Proceedings*, volume 7734 of *Lecture Notes in Computer Science* (pp. 307–318). Berlin: Springer.
van Benthem, J., & Velázquez-Quesada, F. R. (2010). The Dynamics of Awareness. *Synthese*, *177*(Suppl.-1), 5–27.
Velázquez-Quesada, F. R. (2013). Explicit and implicit knowledge in neighbourhood models. In D. Grossi, O. Roy, & H. Huang, (Eds.), *Logic, Rationality, and Interaction—4th International Workshop, LORI 2013, Hangzhou, China, October 9–12, 2013, Proceedings*, volume 8196 of *Lecture Notes in Computer Science* (pp. 239–252). Berlin: Springer.

Chapter 4
Justification as a Dynamic Concept

Abstract This chapter is focused on the concept of 'justification', which I will analyse taking into account various perspectives. Firstly, I will define the notion by itself and review its applications and uses in mainstream Epistemology. Afterwards, I will examine how the concept has been used in Justification Logics and some of its contemporary developments. Eventually, I will present a different logical understanding of the concept; and finally, introduce my interpretation of 'justification' as a process and embed it in the EAK-Schema.

Keywords Justification · Justified true belief · Justification logic · Observation · Deductive inference

4.1 The Role of Justification in Epistemology

Though much has been already said about the different epistemological approaches to the justification of true beliefs (or information), I have not specified yet what this 'justification' really means. Therefore, in this section I will address what *justification* is, or how it has been understood in different views, and also how it is interpreted in Awareness Justification Internalism (AJI).

What 'justification' is. As already done in previous sections, I will start with the basic definitions[1] from two authoritative dictionaries. The *Oxford Dictionary* defines 'justification' as "1. The action of showing something to be right or reasonable. 1.1. Good reason for something that exists or has been done."[2] While the *Merriam-Webster Dictionary* states that it is "1.a: the act or an instance of justifying something: vindication. b: an acceptable reason for doing something: something that justifies an act or way of behaving."[3]

From the above definitions one can conclude that, in some vague sense, 'justification' is a process ("action", "act" or "reason for") that results in a 'justified object'.

[1] I disregard the considerations about its legal and theological meaning, for simplicity reasons.

[2] In https://www.lexico.com/definition/justification, accessed 04/22/2020.

[3] In https://www.merriam-webster.com/dictionary/justification, accessed 04/22/2020.

C. Fernández-Fernández, *Awareness in Logic and Epistemology*, Logic, Epistemology, and the Unity of Science 52, https://doi.org/10.1007/978-3-030-69606-1_4

Thus, after the justification is completed, the object of justification will be in itself what it is. I do not intend to enter any metaphysical debate implying hereby that an object is not complete unless it is justified, but I will state the contrary: if an 'object' (used here in a wide sense) needs to be justified, it will only be 'complete' or 'in itself' after the needed justification has been provided.

In the above description I tried to follow the level of generalisation or abstraction from the definitions in the mentioned dictionaries and therefore used the term 'object'. Once one focuses on the subject matter at stake, namely epistemological justification of knowledge, then the above description should be much clearer. Justification will be *the process through which some true information becomes knowledge*.

As it happened in the case of 'awareness' and 'knowledge', 'justification' is also a very polysemic term, which is applied to a long list of research topics both in common language and academic language (legal, theological, etc.). A complex decision, a vital choice or the week's menu of a household are actions subject to justification, that is, subject to be asked 'why?'. But, I will not consider these types of justifications, since they are neither related to the epistemological nor to the logical aspect of information. Though interrelated, such justifications for any type of actions depend on a longer and different list of factors that have no space or relevance in this research.

Hence, what I will mean with *justification* in what follows is the way in which the 'true belief' becomes knowledge, i.e. the manners in which an agent manages to 'explain' or 'provide reasons for' what she knows. It is, thus, always an *epistemological* or *epistemic justification*, since it refers alone to knowledge. The different interpretations of this process give rise to the different epistemological views about the 'origin of knowledge' (internalism vs. externalism).

Another the way to put it is related with the interrogative adverb 'why'. If an agent states that she knows something, it follows straightforward that she might be asked *why she knows what she knows?* And her answer to this question will contain her justification for her knowledge. In this sense, a justification is always an answer to a why-question containing a rational explanation of what is (presumably) known.

The different types of answers to this question will determine the view one adopts on epistemic justifications and on the problem of the possibility of the knowledge itself. As already explained, knowledge has been traditionally related with the JTB-view (justified true belief). For a belief to be known, it must be justified. The origin of the different forms of justifications there are will determine their theoretical strength or even their plausibility. Broadly speaking, if the origin of justification is internal to the agent, it will give rise to an internalist view; while if the origin is external, an externalist view will follow.

Arguments for Awareness Justification Internalism. The internalist view I will adopt for building the bridge to my interpretation of a dynamic awareness logic is Awareness Justification Internalism (AJI). In the classification mentioned by Pappas (2017), explained in Sect. 1.2, AJI is based on the *Awareness Requirement*, which makes the subject (or agent) aware of something else that "contributes to the justification of belief *B*". Two relevant things follow from this definition: (a) the belief itself is not sufficient for knowledge, i.e. there has to be some *other* information that supports this belief; and (b) this other information is *necessary* for producing the needed justification.

These two facts also imply that this view can be ascribed to a form of *foundationalism*, being this the view that requires that there is one 'basic justifier' upon which the belief is justified. It also leads to the conclusion that this justifier is formed by the information the agent is aware-of. In Sect. 2.1 I addressed the question of which awareness is suitable for being included in AJI and I left one open question for being solved here.

Foundationalism is one of the possible solutions to a traditional problem in Epistemology, the *infinite regress problem*, that consist in the following dilemma: if a belief B_1 is justified on the basis of another belief B_2, this B_2 in turn should be justified on the basis of some B_3, and this in turn on the basis of another. This argument can be continued *ad infinitum*. There have been different answers to this dilemma that can be classified within the following three views: (i) Scepticism: since this chain of justifying beliefs cannot be followed infinitely, knowledge is simply impossible. (ii) Foundationalism: for the chain of justifying beliefs to end, there has to be one basic belief that does not need to be justified in this sense. And (iii) Coherentism: the conclusive fact that the set of justifying beliefs need to fulfil is that it is coherent with itself, i.e. it can form a complex 'web' of reasons that supports the belief.

If I consider the awareness that AJI is mentioning to be awareness-that, which, as should be recalled, is already some type of 'knowledge', then I will have reproduced this regress problem with no convincing explanation to escape from it and it will represent a vicious circle where knowledge is defined on the basis of another type of knowledge. But, if I instead make AJI claim that the agent needs to be *aware-of* the additional information that justifies her belief, then it may seem as I will claim that each type of perceived information is able to become a justifier. To prevent this, I need to follow the foundationalist view and state that the information the agent is aware-of and serves as a 'justification' for her belief is necessary for this belief to be justified.

This might still not seem convincing, since what the agent is aware-of are propositions that she is entertaining at this very moment and whose truth she might not even have considered yet. But there is an important presupposition that is implicit in the epistemological views I am reviewing and that I will not assume in my theoretical framework, namely, that the representation of knowledge, belief and awareness is *static*. I will ascribe the EAK-Schema to the paradigm of *dynamics of information* where the main concept is the 'informational change'. This implies that a given information (or proposition) can start by being part of the awareness-of and then turn into aware-that information, by the application of an epistemic action, that can involve some other information that will be *deductively* or *observationally* related.

Dynamic interpretation of 'justification'. Since the representation of information is now dynamic, what has changed are the different informational attitudes the agent will have regarding the information. As such, she might be aware of some φ without being able to acknowledge the truth of this formula. After she, for example, becomes aware of another information, say ψ, she might have gained enough information to perform an inference and come to the truth of φ, being now φ also part of her awareness-that and having, thus, turned into explicit aware knowledge.

The dynamic interpretation of information leads, thereby, to a different understanding not only of the concept of 'justification', to be addressed bellow, but also of the knowledge forming process. In mainstream Epistemology it seems to be necessary that the information an agent adduces for justifying her belief is completely 'detached' from the belief. In externalist views this information is an external evidence and in internalist approaches it is a different belief that needs to be related to the belief that will be justified. It seems that what matters is the information itself, rather than the informational attitude the agent has towards it. What I try to emphasise from the dynamic view is that, though the information matters, what matters most is the attitude the agent adopts towards it and how this attitude will change thanks to different factors (external and internal).

Being this so, the relevant information that will act as a 'justifier', in the sense AJI presents it, will be the propositions the agent has at her disposal (her awareness-of). Furthermore, the process of justification will be given by the capability of the agent, based on the information she has access to, to perform the needed epistemic actions in order to gain Explicit Aware Knowledge. Then, the content of her justification will be the different actions she has performed and the consequent changes in the informational model.

Since the focus is now on the actions, rather than on the information itself, what will be necessary and relevant is the 'relation' between the information to be justified and the justifying information. Resuming the example above, if she needs to justify φ, of which she is aware-of but whose truth she has not acknowledged yet, she will need to become aware of ψ (related to φ) in order to be able to infer the truth of φ, being then aware-that φ is the case and aware-of φ, it is possible to conclude that she has Explicit Aware Knowledge of φ. Along this process φ has not changed, what has changed is the informational attitude she has towards this information and her gain of new information that allows the inference to the truth of φ.

The relation between these two pieces of information can come in two different forms, depending on the action that she performs: (i) there can be a *deductive relation*, meaning that she infers the truth of φ on the basis of ψ performing an action of 'deductive inference'; or (ii) there can be an *observational relation*, meaning that she observes ψ (in the sense that she can acquire it from a trustful source) and through this action of 'observation' she becomes aware-that it is the case.

4.2 Justification Logic and a Different Approach

In this section I will review Justification Logic, an extension of Epistemic Logic that has been developed in the past decades. I will analyse the understanding of the concept of 'justification' in this logic and mention some interesting contemporary proposals that combine the notion of 'justification' with related notions from

Dynamic Epistemic Logics. Afterwards, I will present an alternative understanding of the concept, that will support the interpretation in the EAK-Schema.

Basics from Justification Logics. As already mentioned in Sect. 3.2 regarding the different strategies for distinguishing explicit from implicit knowledge, Justification Logic (JL) is a branch of logics that models 'justifications' by bringing together Epistemic Logic (EL) and Logic of Proofs (LP). In Artemov and Fitting (2016) the authors define 'justifications' in the following way:

> Justifications are formal terms, built up from constants and variables using various operation symbols. Constants represent justifications for commonly accepted truths–typically axioms. Variables denote unspecified justifications. (Artemov and Fitting 2016, para. 1.1).

The equivalence between 'justifications' and 'formal terms', as well as the fact that JL is inspired in LP, implies that it is mathematically oriented. As such, it adopts a formal structure that, at least at first sight, has little in common with the epistemological interpretation of knowledge. Interestingly, and probably to escape from these type of allegations, the authors start the description of the origin of this logic alluding to the JTB-view and announcing that JL will finally include the 'J' (justification), which was absent in EL (that only considered the 'true belief'). Therefore, this logic intends to cover what EL could not.

JL takes the standard structure from EL, with its implicit logical omniscience, and extends it with components from the Logic of Proof (LP), understanding now 'proofs' as 'justifications'. The language of JL includes *justification terms* (t), such that if F is a formula, and t is a 'justification term' for F, than $t : F$ is also a formula or an 'epistemic assertion', to be read as *t is a justification for F*. The authors clarify the following:

> Justification Logic does not directly analyze what it means for t to justify F beyond the format $t : F$, but rather attempts to characterize this relation axiomatically. (Artemov and Fitting 2016, para. 2.1).

It follows from this quote that this approach, as they state in the introductory section, takes a purely formal point of view and, though not always, most of the time it continues to represent idealised agents. Albeit the agents of JL will be able to distinguish explicit from implicit knowledge, as well as the information that acts as justification terms, the bigger picture still shows agents that are closer to 'perfect reasoners' than to human beings. Thus, from a theoretical point of view, JL can be seen as another possible solution to the problem of logical omniscience that only solves it partially.

From a purely logical point of view, conversely, JL represents a very inspiring and promising option, incorporating the concept of 'justification' into the object-language of their system. One of its main notions are the *operations* on justifications, meaning that this logic not only defines 'justification terms', but also different forms of combining these terms. The operation of *application* ($\cdot$) creates a more complex term based on two previous terms. If t and s are justification terms and an operation of 'application' is executed, then $[t \cdot s]$ will be a justification term. The application operation is applied to a Modus Ponens inference: if $s : (F \rightarrow G)$ and $t : F$ then

$[s \cdot t] : G$. In less formal terms, the justification terms can be combined in order to serve as a justification of an inference step.

The other relevant operation they present is the *sum* $(+)$, which adds another justification term to an already existing one. If t is a justification term for F such that $t : F$, then s can be added in the sense that it now widens the content of the justification, being now $[t + s] : F$. Informally, if a formula is justified on the basis of a concrete evidence, there can be another evidence that extends the justification.

The semantic model from EL is extended with an *evidence function*, that maps the justification terms and the formulas they justify to a set of worlds. Then, an epistemic assertion $t : F$ will be true at a given world if both F is 'believable' at this world (implicitly known in the sense of EL) and t constitutes a 'relevant evidence' for that formula in the given world. Being this so, the authors note that "[i]t is important to realize that, in this semantics, one might not believe something for a particular reason at a world either because it is simply not believable, or because it is but the reason is not appropriate" (Artemov and Fitting 2016, para. 3.1).

As mentioned in the previous chapter, the explicit information corresponds to the 'justified information', while the implicit remains untouched regarding the standard approach in EL. The fact that the semantic clause requires the information to be known ('believable' in this case) as well as to be justified, resembles the structure from Awareness Logic, where the explicit information needs to be implicitly known and also belong to the awareness of the agent. Both approaches take the structure from EL and extend it with a different concept. In the case of JL the concept is more complex, since what is interpreted as a 'justification' can be formed by different components and is inspired in the concept of 'mathematical proof'.

This approach has given rise to a whole branch of logics, when extended with other concepts or when the mathematical components of the semantic structure are changed. What is clear, is that it is a formal interpretation and as such often looses the intended link to the epistemological interpretation of knowledge. Artemov and Nogina start their paper stating the assumptions this logic has to make:

> Epistemic systems with justification based on the logic of proofs LP use the following plausible assumptions: (1) each axiom has justification; (2) justification is checkable; (3) justification assertion of a statement implies knowledge of this statement; (4) any justification is compatible with any other justification. (Artemov and Nogina 2005, p. 279).

From the formulation itself of these assumptions, one can see that they are purely and formally oriented (for example, the use of "checkable" implies a mathematical or computational point of view). When interpreted strictly from an *epistemological point of view*, probably the only 'plausible' assumption will be (3), since if an agent has a justification for an information this implies that she knows it. Regarding the other three, I do not think that they could be easily applied to a human subject and her justified knowledge. (1) The mere concept of 'axiom' is complicated to be related with an epistemological notion, unless one understands a 'basic justifier', some information she has acquired as evidence, as an 'axiom'. Then, (2) the fact that a justification can be checked[4] seems to allude to its internal structure, when one thinks of justifications as evidences that support ones belief, these evidences do not need to be checked in any theoretical sense. If this concept is interpreted regarding introspection, then it would just imply that the agent is capable of positive (and maybe also negative) introspection, meaning that she can know what she knows and on the basis of what she knows it. Finally, (4) the compatibility between justifications will be given, in an epistemological sense, by a set of factors that include the context, the rest of the information the subject bears in mind and the specific information she has to justify. In this sense I do not think that 'any' justification will be suitable for being combined with any another.

This argument may serve to conclude that from the theoretical perspective I am adopting, and with the ultimate goal of tending a bridge between the epistemological view of AJI and the theoretical logical framework, Justification Logic, at least in its first presentations, does not fit the EAK-Schema. Furthermore, regarding the concept of 'justification', which is the topic of this chapter, JL represents it in a mathematical style, but does provide little more information to *what* a justification is. Nevertheless, JL has been extended in many ways and I shall review at least one of these extensions before presenting another theoretical interpretation of the concept of 'justification'.

Developments of Justification Logic. JL has inspired a long list of interesting, and sometimes very complex, proposals for solving the problem of logical omniscience, as well as for creating semantic structures that are able to model the informational attitudes of the epistemic agents in an increasingly more and more fine-grained manner. The tremendous growth of this branch of logics is almost unmanageable, all the more if one does not intend to apply this type of logics.

For all that, my only purpose here is to draw attention, in a schematic way, to some of the contemporary developments that have been published and that can incorporate the notion of 'awareness' into the general structure from JL in some sense (sometimes without alluding directly to it). What I will highlight is the advantage this combination provides and the fact that these approaches either adopt a dynamic perspective or can be extended for adopting it, which helps them to create a more intuitive representation of information.

One of these proposals, which I will take as paradigmatic, is "The logic of justified belief, explicit knowledge, and conclusive evidence" (Baltag et al. 2014), where the authors take elements from Justification Logic, Awareness Logic, as well as from different contemporary developments of Dynamic Epistemic Logic and Public Announcement Logic for "reasoning about a notion of *completely trustworthy* ("*conclusive*") *evidence* and its relations to *justifiable* (implicit) belief and knowledge, as well as to their *explicit justifications*" (Baltag et al. 2014, p. 49. Original Italics).

This approach is of course very broad-ranged, since it covers many aspects of the information an agent possesses and the different forms she might not only access or process this information, but also interact with it and provide complex evidences as justifications. There are many key notions that are relevant to this proposal, but one of

[4]The use of the adjective "checkable" by the authors refers here to its computational meaning, alluding to the verifiability of the justifications.

the most important ones (already explored in previous approaches) is that the content of the 'justification terms' from JL can now be analysed. With the *admissibility relation* they evaluate the connection between a 'justification term' and the formula it justifies. In that way, "*evidence t is admissible for* φ, which intuitively means that: t has the formal structure of a well-formed argument, and, moreover, the structure of an argument in favour of φ" (Baltag et al. 2014, p. 51. Original Italics). With this relation the authors are emphasising that what justifies a believed formula is not a single and isolated piece of evidence, as primitive developments of this concept may suggest, but an *argument* that leads to the believed formula. In some sense they are defining an argumentative process as the 'justification' in itself.

They highlight as a "key concept" the "explicit availability of good evidence". In a nutshell, what they refer to with this concept is that not only do the evidences need to be admissible for the formula they justify, as explained above, but the agent needs to have "actually observed [...] all the basic pieces of evidences used as premises in the argument t", and the agent needs to have "actually construed the argument t" (Baltag et al. 2014, p. 53). All these requirements strike me not only relevant but also very intuitive from a theoretical point of view, which is one of the goals of this proposal.

Though I might agree with most parts of this approach I find it too complex for my purpose here. Of course, the more fine-grained one wants to be, the more complex the semantic and syntactic structure of the logical system will be. There is a huge list of technically complex proposals that end up being very fine-grained; and sometimes even too fine-grained, drawing distinctions that human subjects do not even perceive (though computing machines probably will). My goal lies far from this type of proposals, since I am interested in presenting a general picture with a small list of basic concepts, that when combined with the appropriate epistemic actions will show a general theoretical schema of the informational attitudes that an epistemic agent may have.

A different approach of Justification. After the review of JL and a brief description of what can be obtained through its developments, I want to turn my attention to a different form of understanding the concept of 'justification' inspired by the Aristotelian conception of the knowledge-forming process.

So far, I have reviewed the notion of 'justification' in Epistemology and in its most prominent application in contemporary logics, i.e. Justification Logic. But, I started this chapter stating that, to me, 'justification' is a process, rather than something tangible. In other words, I will regard 'justification' strictly as a concept to be defined in the meta-language, instead of incorporating it into the object-language, as is done in JL. The content (justifiers) through which a justification is performed is what most approaches have taken to be synonymous with 'justification', but I disagree.

In what follows I will maintain that the 'justification' in itself is a *process* through which the agent (or human subject) justifies her information, meaning that she provides reasons that support the information she claims to know. The only informational attitude that is subject to be justified is 'knowledge', in the EAK-Schema it is 'explicit aware knowledge'. The content of this process of justification will be given by the epistemic actions, as will be detailed in the next section. When Aristotle describes the different sciences he considers and the possibility of knowledge, he alludes to the processes of gaining knowledge, or, in contemporary terms, the processes of 'justification'.

In Giordani (2013), a paper that introduces into JL the concept of 'truthmakers' for describing justifications of both implicit and explicit information, the author mentions his inspiration in the Aristotelian view. He says that "[a]ccording to Aristotle's seminal analysis, knowledge can be acquired by either *evidential* or *inferential* processes" (Giordani 2013, p. 323. Original Italics). With this reference, he draws a distinction between a direct (evidential) and an indirect (inferential) justification, with the former being related to the intuition of the agent and the latter connected to the deductive capabilities.

What is worth highlighting from this proposal is the role that the 'truthmakers' will play, since the author uses this concept for being able to distinguish different types of justifications and different connections between the epistemic agent, her evidences and the relation to the truth of these evidences. He states that "a truthmaker for a proposition is any actual state of affairs that makes such proposition true" (Giordani 2013, p. 328). He formalises this concept with variables representing the "state of affairs".

Though again, this approach lies far beyond my goals, it is interesting to notice that Giordani has a background intuition of 'justification' as a process through which an agent establishes a certain connection to the actual world that helps her form the truth of certain information from which she will be able to justify, directly or indirectly, her knowledge. As in the preceding approach, the 'access' of the agent to the information (understood as 'awareness' in a broad sense) is a relevant concept, specifically the access to the truthmaker that makes a given information true and, thus, justified. As the author points out in the introductory pages, he follows the JTB-view, but he does not commit himself to neither internalism nor externalism.

Resuming my interpretation of 'justification' as a process and taking the general background intuition from Giordani (2013), I will claim that justification is a process where the information that the agent employs as a justifier can come from two different sources: either from an action of observation, through which she will gain new and true information (similar to the evidential process); or from an action of deductive inference, where she performs an inference step with the true information she already possesses and obtains, thus, as a conclusion, some new information that will be justified (similar to the inferential process). The Aristotelian inspiration can serve here as a foundation of the dynamic interpretation of justification I will maintain.

Before focusing on the dynamic concept of justification and the role of epistemic actions, I want to draw attention to a paper that mentions the same notions I will employ in what follows. In "Justifications, Awareness and Epistemic Dynamics" (Sedlár 2013), the author introduces 'awareness' into a multi-agent justification logic. He states that "justification terms describe *epistemic actions* within specific *groups of agents*. As such, these actions result in alternations of awareness and explicit knowledge of the agents in the group" (Sedlár 2013, p. 307. Original Italics).

This means that the justification formulas he introduces will describe the result of performing some epistemic action. Inspired by the fact that both JL and Awareness Logic (AL) have a similar structure, he mixes them in a way that now the operations between justification terms serve as connections between the aware information of the agents. Forming a complex justification term, due to an operation, can be interpreted as having performed an epistemic action that extends the awareness of the group, for example.

I agree with the background intuition of placing 'justification' in the epistemic actions, while, again, since this proposal stems from JL and AL, it takes implicit knowledge as the primary notion and I do not agree with this choice, as explained in Sect. 3.3. What is worth noticing, is that Sedlár provides a possible formal model for the theoretical notions I will maintain, with its key proposal being the combination of justifications and epistemic actions.

4.3 A Dynamic Interpretation of Justification

To conclude this chapter I am going to present the proposal of the concept of 'justification' in the EAK-Schema. I will begin by explaining the object of justification (Explicit Aware Knowledge), then, define the role that justification will play in the EAK-Schema, and finally, describe the two relevant actions, that intervene in the process of justifying some information.

Explicit Aware Knowledge: object of justification. In Sect. 3.3 I defined this concept as the information an agent is considering at a given moment and whose truth she has acknowledged. This implies that for the agent to know something explicitly she needs to be aware-of it and aware-that it is the case.

The combination of both *awareness of* and *awareness that* already entails the sense of 'justification' I am defining here. If the agent is aware of a given φ, she is able to use this information and create new information based on it. Once she is also aware that it is the case, she has acquired the truth of this φ. There can be information she is aware-that, but is not considering right now, that is the reason for explicit knowledge to require the awareness-of component. Considering the justification of the true information, it follows straightforward that the process that brings her to be both aware-of and aware-that is what constitutes the justification.

Dynamic justification for Explicit Aware Knowledge. As such, justification is a process that ends whenever a given information is part of the awareness-of and the awareness-that of the agent. There are, though, two implicit requirements I have mentioned in the previous section and that are mandatory in a simple and less fine-grained approach like the one I will introduce. These requirements are: (a) the *truth* of the aware-of information that serves as the content of the justification; and (b) the *connection* between the content of the justificatory information and its object, i.e. the connection between the aware-of and the aware-that information.

Regarding (*a*) there is little more to say, most logical approaches have similar requirements. For example, in PAL (Public Announcement Logic) what is being announced needs to be true, it is included in the model of PAL as a 'precondition'. When building the bridge to a human subject and the corresponding epistemological view, it is required that the information this subject applies in the process of justification is true. Regarding (*b*), the connection or relation between the aware-of information and the known information, there are two main connections: a *deductive* and an *observational* or *evidential* one, each of them depends on one of the two epistemic actions the agent will perform, namely 'deductive inference' in the former case and 'observation' in the latter case. This implies that the connection will be given by the informational attitude of the agent towards this information, both in a moment preceding the performance of the action and in a subsequent moment, as explained below.

The role of Dynamic Epistemic Actions. As such, the changes in some parts of the departing epistemic model that the actions provoke, and that are captured by the resulting model, will determine the inter-connection between the information. I will present five actions that represent the different changes in the informational attitudes of the agent that my framework will be able to represent. Yet, only two of these actions will be relevant regarding the justification: deductive inference and observation.

On the one hand, an action of **deductive inference** will represent a standard modus ponens step inside the aware information. If the agent knows explicitly already $\varphi \rightarrow \psi$ and φ (meaning she is both aware-of and aware-that), then an action of deductive inference will make her deduce ψ as a logical consequence of what she already knew, being now also ψ explicitly known. This is the form of 'internally' expanding the explicit knowledge, since it is based on her own information, upon which, only via deductive reasoning, she is able to create new information. In this case, the content of the justification is the entire inference, which, of course, can be formed by more complex formulas or by a whole chain of inference steps.

On the other hand, the action of **observation**, inspired by the 'public announcement', but taken here in a single-agent scenario, provides directly an explicitly aware knowledge. With 'observation' one can understand different situations where the agent obtains, from an external source, a true information. These sources can be, for example, experts on the topic at stake or an academic source that provides true information. What is relevant here is that the source needs to be not only truthful, but also trustful. Once the agent has obtained some information by observation, the source and the channel of information will constitute her justification.

Henceforth, the justification itself, will lie in the capability of the agent of performing these two actions, and the content of the justification will be theoretically given by a 'reconstruction' of the content of the different actions the agent has performed for obtaining Explicit Aware Knowledge of some information.

References

Artemov, S., & Fitting, M. (2016). Justification logic. In E. N. Zalta, (Ed.), *The Stanford Encyclopedia of Philosophy*. Metaphysics Research Lab, Stanford University, winter 2016 edition.

Artemov, S. N., & Nogina, E. (2005). Introducing justification into epistemic logic. *Journal of Logic and Computation*, *15*(6), 1059–1073.

Baltag, A., Renne, B., & Smets, S. (2014). The logic of justified belief, explicit knowledge, and conclusive evidence. *Annals of Pure and Applied Logic*, *165*(1), 49–81.

Giordani, A. (2013). A logic of justification and truthmaking. *Review of Symbolic Logic*, *6*(2), 323–342.

Pappas, G. (2017). Internalist versus externalist conceptions of epistemic justification. In E. N. Zalta (Ed.), *The Stanford Encyclopedia of Philosophy*. Metaphysics Research Lab, Stanford University, fall 2017 edition.

Sedlár, I. (2013). Justifications, awareness and epistemic dynamics. In S. N. Artëmov, & A. Nerode, (Eds.), *Logical Foundations of Computer Science, International Symposium, LFCS 2013, San Diego, CA, USA, January 6–8, 2013. Proceedings*, volume 7734 of *Lecture Notes in Computer Science* (pp. 307–318). Berlin: Springer.

Chapter 5
EAK-Schema and Its Epistemic Actions

Abstract This chapter constitutes the core of this research and is structured as follows: Sect 5.1 will describe each of the central notions and afterwards combine them in a diagram that will show how these concepts converge into the EAK-Schema that represents the informational attitudes of the agent, leading to Explicit Aware Knowledge. Then, Sect. 5.2 will describe five different epistemic actions. Each action will move information from one zone of the diagram to another, or, in other words, each action will reflect an informational attitude of the agent and transform the information she has access to. Finally, Sect. 5.3 will enumerate some advantages of the EAK-Schema both regarding Epistemology and Logic.

Keywords Conceptual schema · Explicit Aware Knowledge · Awareness-of · Awareness-that · Epistemic actions

The thematic thread that I have developed so far, and that concludes with this chapter, is twofold: (i) from Epistemology to Logic; and (ii) from Logic to Philosophy of Logic. In its first direction (i) I unravelled the concepts of 'awareness', 'knowledge' and 'justification', respectively Sects. 2.1, 3.1 and 4.1, and provided my reasons for considering one specific interpretation that can serve as a foundation for a dynamic interpretation of Awareness Logic. While doing this, I discarded many views, always keeping this main goal in mind. In a nutshell, I laid the foundations for the conceptual schema I will present below, that will help to clarify many blurred notions which are commonly employed in contemporary Epistemic Logics.

The inverse direction (ii) is exhibited among Sects. 2.2, 3.2 and 4.2. In these sections I have described, from the point of view of Philosophy of Logic, different interpretations and uses of the mentioned main concepts in some logical approaches. Having in mind the theoretical correlation to the epistemological view, I highlighted the pros and cons of the analysed logical approaches, alluding always to what I presented as my proposal: the EAK-Schema. In short, I have picked up from different contemporary proposals the core notions that will intervene in the EAK-Schema, while showing their advantages from a purely theoretical point of view.

In this chapter, which represents the core of this research, I will put all the presented 'pieces' together and finally introduce the complete EAK-Schema. This schema brings the main goal of this research to its end and provides, thus, a theoret-

C. Fernández-Fernández, *Awareness in Logic and Epistemology*,
Logic, Epistemology, and the Unity of Science 52,
https://doi.org/10.1007/978-3-030-69606-1_5

ical framework that will serve both as a correlation to the epistemological view of *Awareness Justification Internalism* (AJI), and as a theoretical basis for a dynamic re-interpretation of explicit knowledge as a combination of awareness-of and awareness-that.

Though every concept has the same relevance, when it comes to describing the informational attitudes of the agent from an epistemological and logical point of view, what matters most is the 'Explicit Aware Knowledge' (EAK), that is, what the agent really knows. As such, though the other attitudes will be represented, they all lead to the EAK and show the process for obtaining it.

5.1 Epistemic Concepts Combined into One Diagram

I will start this section recalling the definitions of the main epistemic concepts I have presented. Afterwards I will combine them in a diagram (the EAK-Schema), signifying first the formation process of the diagram and second the different zones and its corresponding informational attitudes.

Core epistemic concepts. In what follows I will recall what has been presented in Sects. 2.3, 3.3 and 4.3. I will follow the order of the mentioned sections, though they will be re-ordered when presenting the diagram. The two types of awareness as well as the two types of knowledge represent the informational attitudes the agent can have. When describing the diagram I will explain which of them can be combined and how.

- **Awareness of**: the agent being aware of φ means that she entertains φ. Every information that is *available* to her will form the set of aware-of information. This concept can be seen as the language she has at her disposal, i.e. everything upon which she can form more complex formulas. The important feature is that the agent need not be aware-of the truth of this information, as such, she can be aware-of false information, but also of doubts, mere perceptions without truth-value, etc. The set of information she is aware-of can be extended or reduced with the actions of 'becoming aware-of' and 'becoming unaware-of', respectively.
- **Awareness that**: if the agent is aware that φ is the case, this means that she has acknowledged the truth of φ. The acknowledgement of the truth implies that the information she is aware-that is knowledge-like, in the sense that it is true information, but is not justified yet. For the agent to become aware-that some information is indeed the case, she needs to recognise the truth of this information and be able to deliver some 'reasons' for this information to be true.
  The set of information the agent is aware-that it is the case can be increased with the epistemic action of 'deductive inference', through which she will have deduced from what she already 'knew' some new information. The counterpart of this action is 'forgetting', an action that makes the set of aware-that information shrink, since what she forgets is how she obtained the truth of the information.

- **Explicit Aware Knowledge (EAK)**: for an agent to have Explicit Aware Knowledge of some φ she needs to be aware-of φ and aware-that φ is the case. By being aware-that she already has acknowledged the truth of the information, but this requirement alone does not guarantee that it is 'explicit knowledge', since being explicit means that she actually knows it at a given moment. For this to be the case she will also need to be aware-of the information. The combination of both provides automatically the needed justification, as will be mentioned below.
  The relevant epistemic actions that will make an information part of the agent's Explicit Aware Knowledge are 'deductive inference' and 'observation', since both these actions provide the needed justification.
- **Implicit Knowledge**: the *closure under logical consequence* of the awareness-that will be the implicit knowledge of the agent. Since the set of aware-that information represents true information the agent has acknowledged as such, irrespective of her being aware of it or not at a given moment, whatever can be deduced from this set of information will constitute the implicit awareness-that. When the agent is also aware-of this informaiton it will constitute 'implicit knowledge'. This concept is logically meaningful, but it can only be considered a 'potential concept' from an epistemological point of view. Thus, the agent will not 'have' implicit knowledge as such, but it is an important concept when considering the possible inferences the agent may perform in order to extend her awareness-that. Of course, the only epistemic action that can 'create' the implicit knowledge is 'deductive inference'.
- **Justification**: obtaining Explicit Aware Knowledge through some epistemic action constitutes the process of justification. Since explicit knowledge needs to be justified true information and it is already true information thanks to the awareness-that, the justification lies in the requirement for the agent to be aware-of, and in the actions that make the true information part of the awareness-of, i.e. that turn the information into EAK.
  In this way, what can justify Explicit Aware Knowledge are the actions of 'deductive inference' and of 'observation'. The former will imply a *deductive justification*, where the agent has inferred the information she now knows explicitly, while the latter is a form of *observational justification*, where the agent has obtained true information from an external source, that makes her automatically be aware-of and aware-that.

5.1.1 Formation of the Diagram Combining Epistemic Concepts

In what follows I will present step by step a diagram that combines the mentioned epistemic concepts, including an example that illustrates which informational attitudes this concept refers to. I will introduce one concept at each step and analyse the different 'zones' of information that have been created by overlapping some of these concepts.

Starting with Awareness-that (Fig. 5.1).

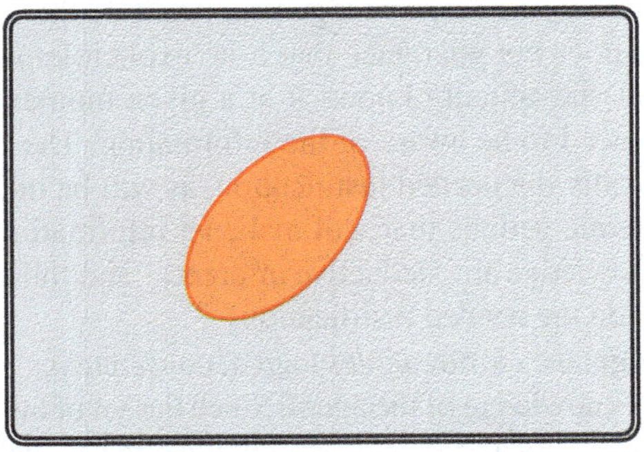

Fig. 5.1 Awareness-that

Following the background intuition from the syntactic approach by Konolige (1984) I will start with the awareness-that of the agent. *Every* information whose truth she has acknowledged is pictured with this orange ellipse. I recall that it is every true information, irrespective of the moment of evaluation.

The grey rectangle represents the domain, that is, every information there is. What is included in the orange ellipse is exclusively those formulas she 'knows' to be true at any moment. Since *awareness-that* represents already a type of knowledge, in some examples I will use the verb 'to know' as an informal synonym for 'being aware-that some information is the case', in order to simplify the language and make the examples more readable.

Example 5.1.1 Olivia is *aware that* π is an irrational number.

Deducing Implicit Awareness-that (Fig. 5.2).

Fig. 5.2 Implicit Awareness-that

Based on the information the agent is aware-that it is the case one can form the set of implicit awareness that. The green ellipse (which includes the orange one) represents here everything that can be deduced from whatever the agent is aware that it is the case. The fact that the implicit information is what can be deduced from the 'really known' information corresponds to the common sense intuitions regarding human beings.

The green ellipse can also be seen as the *closure under logical consequence* of the orange one, since it is formed by every information that is 'reachable' via deductive reasoning on the basis of the aware-that information. As such, it represents an idealised or omniscient agent who has exhausted all possible inferences there are. Being 'implicit' means that it is always a *potential* information that is taken as an ideal objective only to be reached by a 'perfect reasoner'.

Example 5.1.2 Olivia is *implicitly aware that* π is not a rational number.
Her being aware that π is an irrational number (Example 5.1.1) implies that π is not a rational number, since this numerical classification is exclusive.

Incorporating Awareness-of (Fig. 5.3).

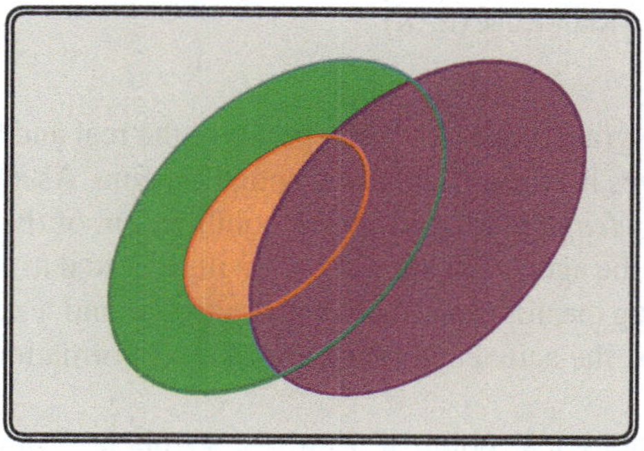

Fig. 5.3 Awareness-of

The key notion in this schema, taken from Fagin and Halpern (1988), is the concept of *awareness of*, that is, what the agent is entertaining at a given moment. The purple ellipse on the right represents all this information. As explained above, the agent can be aware of any type of information and she will not consider its truth value regarding her awareness-of. Every information in the domain is subject to be part of the awareness-of, here I depict a subset of it. Since what is interesting is the combination of this notion with the former two, the purple ellipse needs to intersect the other two ellipses.

The combinations with implicit awareness-that and with the awareness-that will be analysed in what follows, for now I will only focus on the pure awareness of, that is, on the 'language' the agent has at her disposal.

Example 5.1.3 (*a*) Olivia is *aware of* irrational numbers having infinite digits that never repeat. (*b*) Olivia is *aware of* the number $\sqrt{2}$ being irrational.

She is aware of this two statements in the sense that 'she has heard of it', but it is not possible to evaluate the truth of them only regarding her awareness-of, though they are indeed true statements.[1]

Explicit Aware Knowledge as *Awareness-that* and *Awareness-of* (Fig. 5.4).

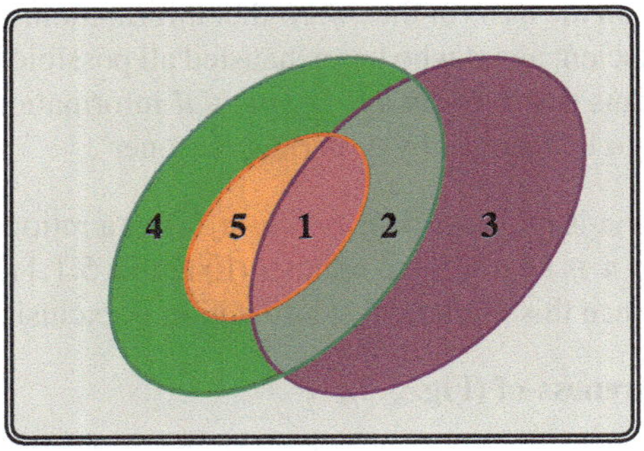

Fig. 5.4 Explicit Aware Knowledge (EAK)

The kernel of the diagram, zone (1), corresponds to the real and actual knowledge of the agent, which gives its name to this conceptual schema. As already described, the *Explicit Aware Knowledge* is formed by the combination of the awareness-that and the awareness-of of the agent. For simplicity, I will refer to it as 'explicit knowledge' in the examples, being the difference between 'knowing' and 'explicitly knowing' the awareness-of, that is, the actual entertainment of the information at a given moment by the agent.

From an epistemological point of view, the Explicit Aware Knowledge is the only part of the diagram that can be equated with the notion of 'knowledge' in Epistemology, since it corresponds to true information the agent has acknowledged (awareness-that) and that will be justified due to her being aware-of it. The remaining parts of the diagram will also be epistemologically relevant, though they stem from the logical point of view. In the diagram, only the part of awareness-that that intersects with the awareness-of is subject to be named 'explicit knowledge' and in there one finds the real knowledge the agent is considering.

Example 5.1.4 (i) Olivia *knows explicitly* that irrational numbers have infinite digits that never repeat. (ii) Olivia *knows explicitly* that $\sqrt{2}$ is an irrational number.
Examples 5.1.3 (a) and (b) showed that Olivia was aware of these statements, but without considering its truth. Now, regarding her awareness-that I state that she has acknowledged the truth of them. Being this so, she certainly knows explicitly these statements.

[1]Notice that this set could also include false information, which would always be located outside the green ellipse.

Implicit Knowledge as *Awareness of* and *Implicit Awareness-that* (Fig. 5.5).

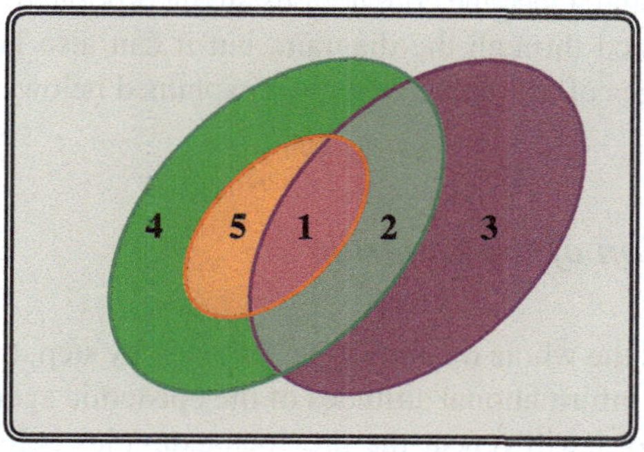

Fig. 5.5 Implicit Knowledge

The concept of *Implicit Knowledge* in itself is a logical construct. In this diagram it is represented as the part of the implicit awareness-that that intersects with the awareness-of of the agent, pictured with number (2). This zone of the diagram is probably the less intuitive regarding its correlation to human beings in an epistemological interpretation. From the logical point of view the implicit knowledge is formed by the information that can be deduced from the Explicit Aware Knowledge. As such, it has to be necessarily inside the scope of the awareness-of.

When considered from an epistemological point of view, this zone corresponds to the 'potential aware awareness-that' of the agent, that is, the information the agent can infer from what she has already acknowledged as true and that she is entertaining, but whose truth she has not acknowledged nor does she have sufficient information yet for acknowledging it.

Example 5.1.5 Olivia *knows implicitly* that $\sqrt{2}$ has infinite digits that never repeat. She has Explicit Aware Knowledge of irrational numbers having infinite digits that never repeat (Example 5.1.4 (i)) and of $\sqrt{2}$ being an irrational number (Example 5.1.4 (ii)). From there, she she has the means for inferring that $\sqrt{2}$ has infinite digits that never repeat.

The diagram does not only serve as a very useful visual aid for the description of the EAK-Schema, but it will also be part of the explanation of the different epistemic actions. I developed this diagrammatic structure inspired by the diagrams included in Konolige (1984) and represented in Sect. 3.2. Konolige employs two different diagrams, one for representing his own syntactic proposal and another for the Awareness Logic from Fagin and Halpern (1988). While I could have used the latter, I considered it more intuitive to take the former and add the awareness-of set as a superposed ellipse that intersects the first two.

This formation process I have developed here corresponds to the intuitive and theoretical order of appearance of the different concepts in the EAK-Schema. Before taking 'awareness' into account, the epistemological theories considered what the subject knew and what she was able to reach with that (implicit). After the incorporation of the *Awareness Requirement*, and thereby the shaping of AJI, the notion of 'awareness' is essential. Once the core notions are incorporated, the epistemological view can be explained through the diagram, but it can also help to represent the informational changes of the agent, as will be explained below.

5.1.2 Description of the Diagram

Having now formed the whole diagram (Fig. 5.6) step by step, the resulting description of the different informational attitudes of the epistemic agent I am representing is the following one. I will repeat the previous examples below each region, and thereby show to which zone of the diagram they correspond.

(1) **Explicit Aware Knowledge**: awareness-that and awareness-of.
5.1.4: (*i*) Olivia *knows explicitly* that irrational numbers have infinite digits that never repeat. (*ii*) Olivia *knows explicitly* that $\sqrt{2}$ is an irrational number.
(2) **Implicit Knowledge**: implicit awareness-that and awareness-of.
5.1.5: Olivia *knows implicitly* that $\sqrt{2}$ has infinite digits that never repeat.
(3) **Awareness of**: every considered information.
5.1.3: (*a*) Olivia is *aware of* irrational numbers having infinite digits that never repeat. (*b*) Olivia is *aware of* the number $\sqrt{2}$ being irrational.
(4) **Implicit Awareness-that**: outside awareness-of.
5.1.2: Olivia is *implicitly aware that* π is not a rational number.
(5) **Awareness-that**: outside awareness-of.
5.1.1: Olivia is *aware that* π is an irrational number.

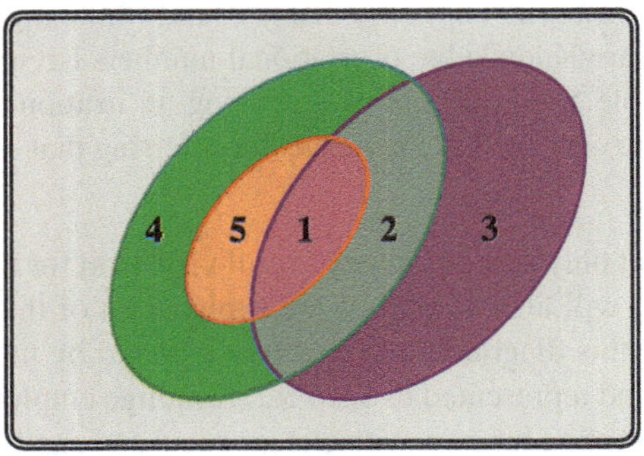

Fig. 5.6 Diagram for the EAK-Schema

The examples corresponding to (3) Awareness-of are repeated in (1) Explicit Aware Knowledge, since for the agent to have Explicit Aware Knowledge she needs to be aware of it and aware that it is the case. Regarding the zones outside the awareness-of, that is, what she is currently not entertaining, the information contained there could be accessible to her with the actions of becoming aware-of and performing a deductive inference, as will be shown in the next section.

So far, the diagram in the EAK-Schema shows a static description of the informational attitudes. But, I intend to adopt a *dynamic perspective* that helps represent the changes in the information the agent possesses. As such, in what follows I will describe five different epistemic actions that represent these changes.

5.2 Changes in the Informational Attitudes: Epistemic Actions

In Sect. 1.1 I introduced the concept of *epistemic actions* and explained their main features. It should be recalled that an epistemic action represents a change in the information the agent possesses, in such a way that either there is an information loss or win, both these cases can include all sorts of side effects that an alteration in the informational state may bring with it.

Formally, the epistemic actions always change some part of the epistemic model by extending or shrinking one or more specific parts. In the EAK-Schema there are two main concepts that are subject to change: *awareness-of* and *awareness-that*, which also constitute the core notions of this understanding. The actions will be represented by changing the amount of the information contained in the corresponding zones of the diagram.

In what follows I will present five epistemic actions that correspond to five intuitively common changes in the agent's information.[2] The first two actions, *becoming aware-of* and *becoming unaware-of* will be related to the changes in the awareness-of of the agent, that is, changes in the attention she puts on all the information there is. A change in the information she entertains can have different causes, just like it can imply diverse effects regarding the remaining information.

The next actions will affect the awareness-that of the agent, that is, these actions will change the acknowledgement of the truth of some information. The action of *deductive inference* represents an inference step the agent performs based on the information she is already aware-that; the counterpart will be the action of *forgetting*, where the agent will 'forget' the truth (how she came to the truth) of a given information, while remaining aware of it.

[2] A preliminary approach to what will be explained here has been published in Fernández-Fernández and Velázquez-Quesada (2018).

The last action is one that performs changes in both awareness-of and awareness-that, namely the action of *observation*. Inspired by the 'public announcement', this action represents a situation where an agent obtains true and aware information from an external source, such that she will automatically explicitly know it.

One of the most important features of epistemic actions is that they prevent the agent from being omniscient, in the sense that they help to represent the different steps an agent performs for obtaining new information. They also prevent the agent from being completely ignorant, because the epistemic actions allow to incorporate new data for increasing her available information.

In what follows I will explain each of this actions in the EAK-Schema with the visual help of the diagram in Fig. 5.6, using arrows that will represent the direction of the information win or loss. I will also provide new examples for each action, describing the informational context.

5.2.1 *The Action of Becoming Aware-Of*

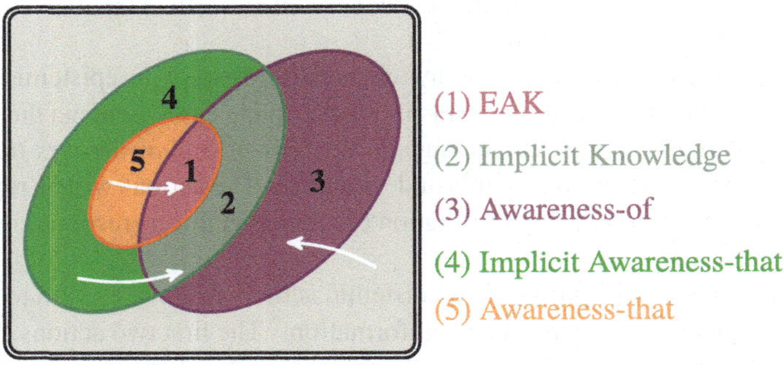

Fig. 5.7 Action: becoming aware-of

Since awareness-of overlaps the other two sets of information (awareness-that and implicit awareness-that) there are three different 'zones' inside the Awareness-of. Each of this parts will be formed by information the agent is entertaining, but there are differences among them (Fig. 5.7). I will describe each of them alluding to the three different forms the action of *becoming aware-of* might take place.

(3) Pure Awareness-of: region (3) is formed with all sorts of information, here there can be false information, beliefs, doubts, etc. But only the true information will be considered regarding the epistemic actions, since they stem from a formal point of view and the initial requirement is that the information used for the knowledge-forming process needs to be true.

For expanding this region the information comes from 'outside' of the ellipse, that is, from the whole domain depicted with the grey rectangle. The arrow on the right shows an action where the agent becomes aware of a brand new information.

Example 5.2.1 *Context*: I am having a conversation with some friends. One of them is commenting on his friend Paul and his age, he is 35 years old. I had never heard of Paul, neither of his age, but now I entertain the information that 'Paul is 35 years old'.

Action: I have become aware of Paul being 35 years old.

(2) Awareness-of of implicit awareness-that: region (2) represents information the agent could deduce based on what she has in region (1), i.e. what she explicitly knows. But before being able to deduce it she needs to become aware of it, hence this action brings to her attention the information she needs for being able to deduce it. The arrow representing this action goes from region (4) to region (2), since it corresponds to implicit awareness-that information that will be now part of the agent's language or attention within her implicit knowledge.

Example 5.2.2 *Context*: I am talking to a friend of mine who is asking about my family. He then asks me about my brother's age. Up to that point, I was not even thinking of this topic, nor do I yet have explicit knowledge of it. But, since I do explicitly know the year of my brother's birth, his age is something I can deduce.

Action: I have become aware of the topic of 'my brother's age' and I have the means for deducing it's content from what I explicitly know.

(1) Awareness-of of awareness-that: region (1), as the kernel of the diagram, represents the Explicit Aware Knowledge (EAK). The awareness-of depicted here corresponds to the attention of the agent on an information she 'knows' to be true. As such, an action of becoming aware of an awareness-that information can be considered as an act of 'remembering', since it brings to light some information the agent really knows, but simply was not considering at this moment. The action from region (5) to (1) is thereby an act of extending the attention of the agent to an information she 'knows'.

Example 5.2.3 *Context*: when I am writing a paper on Epistemic Logic I am not considering my brother's age. If, then, I receive a call from an old friend who, among other topics, tells me that he has seen my brother and knows that he is 22 years old, then I automatically become aware of my brother being 22 years old, an information that I now explicitly know, but before I was only aware-that it was the case.

Action: I become aware of my brother being 22 years old.

Actions that involve becoming aware (though some of them were named differently) have been formally represented and studied in Hill (2010)), van Ditmarsch and French (2011) and van Benthem and Velázquez-Quesada (2010),[3] among others. The above discussion can serve as their theoretical correlate.

---

[3]This paper has been analysed in Sects. 2.2 and 3.2.

5.2.2 The Action of Becoming Unaware-Of

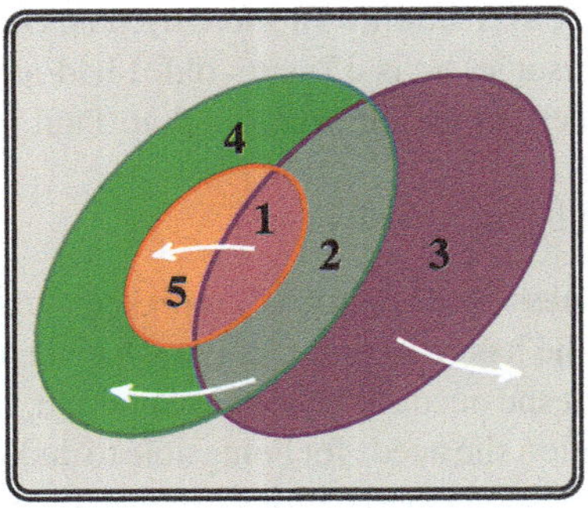

Fig. 5.8 Action: becoming unaware-of

As already mentioned, the action of *becoming unaware-of* is the opposite to becoming aware-of. As such, if regarding the previous action, all of the three possibilities presented acts of focusing on some information or bringing it to light, here I will consider acts of 'not focusing' or 'dismissing' some information (Fig. 5.8). Since an epistemic agent has a limited set of awareness-of, i.e. is limited regarding the focus of her attention, there are different situations where she will disregard some information she is not needing at this moment.

Being the becoming unaware-of action the complementary move regarding becoming aware-of, I will describe the inverse direction of the three types of becoming aware-of actions I mentioned above. Depending on the type of information she is considering, that is, depending on the region of the diagram it corresponds to, by becoming unaware of it, the information will return to the region where it came from. I will continue the examples of the previous action (Sect. 5.2.1) showing now the inverse situations.

(3) Unaware of Awareness-of: the information represented in region (3) corresponds to propositions the agent only has had a first contact with, that is, it represents 'brand new' information. As such, the move from region (3) back to outside the diagram means that she stops considering it, probably because of lack of relevance to another topic she is entertaining at this moment.

Example 5.2.4 *Context*: I had become aware of Paul being 35 years old and of him being friends with a friend of mine, but I did never actually meet Paul. When the conversation with my friends is over, I go back to the office and resume my work on the historical origins of classical logic. At that moment, the information regarding Paul's age has left the topic of my attention.

Action: I became unaware of Paul being 35 years old.

(2) Unaware of implicit awareness-that: region (2) depicts the information that is reachable via deductive inference from the explicit knowledge of the agent. As such, this implicit information, is related to the awareness-that, that is, to what the agent has acknowledged as true. For some of this information in region (2) to leave the awareness-of region and go back to region (4), the possibility of deducing it will also need to be not aware-of. In this way, disregarding some potential information implies that the origin of this potential information also needs to be disregarded, which is usually the case after the agent becomes unaware of whatever she could have deduced from what she knew.

Example 5.2.5 *Context*: After the conversation with a friend of mine, who mentioned the topic of 'my brother's age', a number I can deduce from my knowledge, I go to the grocery store for doing the weekly shopping. While purchasing the different items of my shopping list I completely focus on this task and stop attending the conversation I had with my friend.

Action: I have become unaware of my brother's age.

(1) Unaware of awareness-that: what the agent explicitly knows is also the information she can justify and thus has reasons for it. While the Explicit Aware Knowledge represents the epistemological concept of 'knowledge', it is also the smallest region of the diagram, since it corresponds to the true and justified information the agent is considering at a given moment. The act of becoming unaware of it is probably one of the most common ones, represented informally with the expression of 'I knew it, but I do not remember it now' or 'it is on the tip of my tongue'. This kind of situations, where the agent has no immediate attention on some information she remembers to have known, are the results of becoming unaware-of. A move from region (1) to region (5) therefore represents acts of change of attention.

Example 5.2.6 *Context*: While I was on the phone with an old friend I explicitly knew that my brother is 22 years old. After hanging up and returning to my initial task, writing a paper on Epistemic Logic, the focus of my attention is now directed towards the topic of my paper and not my brother's age. It is still an information I know, but one that I am not taking into account at this moment.

Action: I have become unaware of my brother being 22 years old.

Actions representing 'unawareness' are formally described in van Ditmarsch et al. (2012) and van Benthem and Velázquez-Quesada (2010), for example. As in the previous action, the formal representation of them uses different names, but the background intuition remains. As such, the discussion above can serve as a theoretical correlation to them.

5.2.3 *The Action of Deductive Inference*

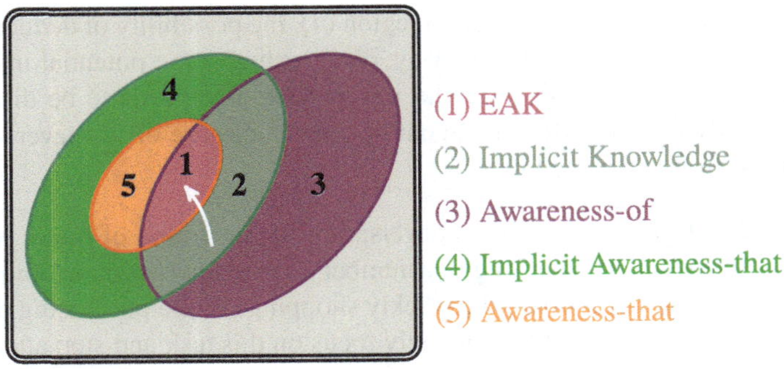

Fig. 5.9 Action: deductive inference

An action of deductive inference represents the rationality of the agent (Fig. 5.9). In the syntactic approaches, like the one presented in Konolige (1984), it is the only possible action for increasing the explicit knowledge of the agent. The move from (2) the **Implicit Knowledge** to (1) the **Explicit Aware Knowledge** in the diagram, represents each step the agent may perform in order to reach omniscience. The agent this diagram depicts is, of course, not logically omniscient, but she is neither completely ignorant. The possibility of performing this deductive step makes her rational.

The general intuition behind this action would be any type of inference, though formally this action will be represented by a Modus Ponens step. With the action of deductive inference the aware awareness-that, i.e. the explicit knowledge, can be extended. There is only one possible direction for this action to take place, since it requires not only a step for turning explicit the implicit information, but it also requires that the agent performs this action while being aware of the involved information. Of course, the complete green ellipse (regions (2) and (4)) represents the implicit awareness-that, like the orange ellipse (regions (1) and (5)) represents the whole awareness-that, but the inference needs to be performed with aware-of information (from region (2) to (1)).

This requirement permits me to state that the *justification* of the explicit knowledge lies in the 'process' of performing the action of deductive inference. If the agent has increased her explicit knowledge on her own, that is, via a deductive inference, then she needs to be aware of how she performed this inference and be able to explain it. This way, the deductive justification, defined in Sect. 4.3, relies on the capability of the agent to perform this specific action.

Example 5.2.7 *Context*: In the situation from Example 5.2.2 I became aware of the topic of 'my brother's age' in a conversation with a friend. Since I explicitly know his year of birth, namely 1998, I can infer from it, with some basic arithmetic, that he is now (in 2020) 22 years old.

Action: I deductively inferred that my brother is 22 years old, since I explicitly know that he was born in April of 1998 and I am aware of this information.

Formal accounts of this action in an Epistemic Logic setting can be found in Ågotnes and Alechina (2007), Duc (1997), Jago (2009) and Velázquez-Quesada (2013). The different approaches represent how an agent can extend her information based on the information she already has. The above discussion serves, as in the previous cases, as a theoretical correlate to what has been done in purely formal accounts.

5.2.4 The Action of Forgetting

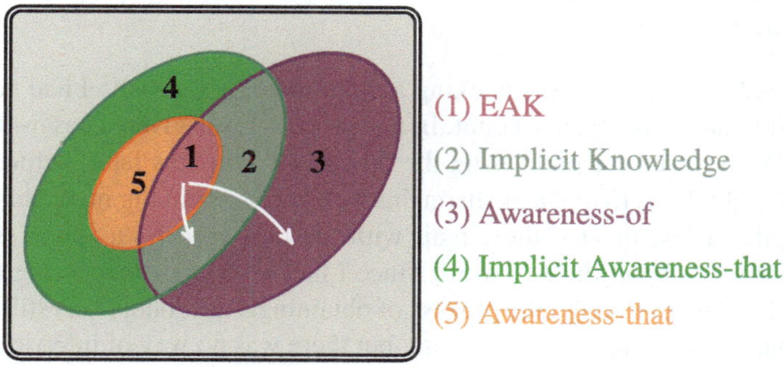

Fig. 5.10 Action: forgetting

The action of forgetting could be seen as the complement to the one of deductive inference, though only in one of its possible performances (Fig. 5.10). When the agent 'forgets' something in the sense that is represented here, what she really forgets is how she came to the truth of the information she was aware-that. In this way, an information that was in region (1) will remain in the awareness-of, but it will go back to the region where it came from. After forgetting an information it can move to region (2) or to region (3). What is common to both types is that it will always shrink the set of explicit knowledge. I will analyse these two cases.

Forgetting: from (1) EAK to (2) Implicit Knowledge: the action of forgetting the deductive step the agent had performed for obtaining her explicit knowledge can be seen as a *simple forgetting*. An information can still be part of the agent's attention (her awareness-of), but she can have forgotten how she inferred the truth of this information, in this sense it will go back to being only implicitly known and a new act of deductive inference will bring it back to explicit knowledge. This type of 'forgetting' is common in cases where the inference steps are quite complicated (like a mathematical theorem and its proof) or the agent does not perform this inference very often (like knowing the age of a baby, commonly expressed in weeks or months, instead of years).

Example 5.2.8 *Context*: I asked my friend about her daughter and she told me that she is already 15 weeks old. From that I inferred that she is almost four months old. Some days later, I have forgotten how many months old her daughter is, but I still know explicitly that she is 15 weeks old, therefore I am able to infer again what I had forgotten.

Action: I forgot that she was almost four months old, based on my explicit knowledge of her being 15 weeks old.

Forgetting: from (1) EAK to (3) Awareness-of: when the agent forgets some explicitly known information and it goes directly back to the pure awareness-of region, then she not only has forgotten how to obtain the truth of it, but also does not have the needed tools for obtaining this truth again on her own. This could be seen as a complicated forgetting, because of the complexity of obtaining again the truth of the forgotten information.

Example 5.2.9 *Context*: I was booking a flight through a website I had never used before and I had to create an account. In this process I was given a password for my account which I used for purchasing the flights. Two months later, before actually taking the flight, I had to log in again on this website for checking in. At this moment I realised that I had deleted the e-mail with the password I was given and I had forgotten how to obtain the password. Since I had no way of recovering this same password I had to go through the process of obtaining a new one. I was still aware-of the fact that I had been given a password, but there was no way of inferring it again.

Action: I forgot how to obtain the password I was given and had no way of recovering it.

Formal accounts of this action have been developed in Fernández-Duque et al. (2015) and van Ditmarsch et al. (2009). It is interesting to note that though it is named 'forgetting' it does not directly correspond to the intuitive sense of the concept, which would be better equated with becoming unaware-of. The interesting feature of the act of forgetting is that what the agent forgets is how she acknowledged the truth of this information, instead of the information in itself.

5.2.5 *The Action of Observation*

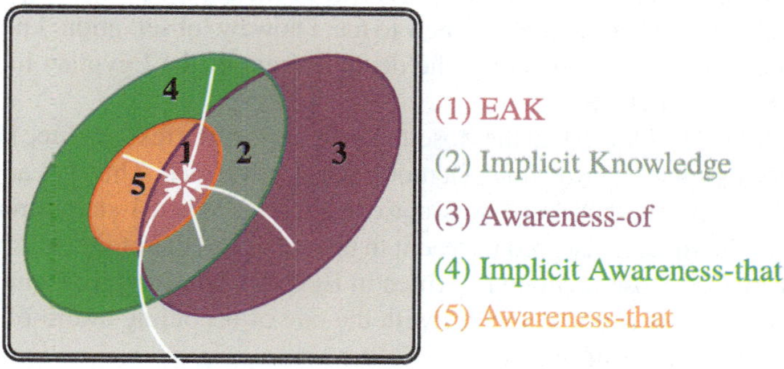

Fig. 5.11 Action: observation

The last action I will present is probably the most intuitive one, as well as the most complete, in the sense that with this only action the agent obtains both awareness-of and awareness-that (Fig. 5.11). An act of observation represents every possibility through which the agent receives true information from an external source. Since what she receives is always true information (already established as a requirement), the information will automatically be explicitly known.

The action of observation is the second possibility for the agent to justify her explicit knowledge. The process of receiving the information and her trusting the source will suffice for an *observational justification*. Of course, this implies that the given source is always truthful and will never pass on to her false information. This requirement is needed for preventing 'strange' cases, like the examples proposed by Gettier.

Regarding the diagram, as showed in the above figure, the information can move from any region to the kernel, EAK. This implies that the agent can observe completely new information (from outside the diagram), information she was simply entertaining (from 3), information she could have reached by deductive inference (from 2), information she already knew but was not considering (from 5) and also information she only could have acquired after first becoming aware of it and then performing a deductive inference (from 4).

Many of the examples I exhibited regarding the previous actions already dealt with an external source (some friend) passing on some information to me, but this information was not always enough for being part of my explicit knowledge. In what follows I will present some new examples where the observed information follows the structure of the action of observation.

Example 5.2.10 Descriptions of acts of *observation* from the different regions in the diagram to Explicit Aware Knowledge:

(a) From outside to (1): I went to the British Museum, saw the Rosetta Stone and bought a book by John Ray explaining its history. Although I already knew some parts of the history of the rock, I did not know that Thomas Young (1773–1829) was the one who started the decipherment of the Egyptian hieroglyphs. Since this information was completely new to me, I now by 'observation' know explicitly that Thomas Young started the decipherment of the Egyptian hieroglyphs through the Rosetta Stone.
(b) From (5) to (1): I knew that the Rosetta Stone has written texts in three languages, namely Egyptian hieroglyphs, Demotic script and Greek. This was an information I was aware-that, but I was not considering. When I started reading the book of the stone I observed it (read it in this case) and therefore I now explicitly know that the Rosetta Stone is written in Egyptian hieroglyphs, Demotic script and Greek. This action coincides with the one of becoming aware-of, which is included in the act of observation.
(c) From (4) to (1): The reading of the book also taught me that the Demotic script is a form of hieroglyphic abbreviation that was used for writing daily matters in Ancient Egypt. By reading this I automatically have explicit knowledge of it, but since I knew that there was a Demotic script related to the Egyptian hieroglyphs, it is an information I could have deduced after becoming aware of the fact that the stone was also written in Demotic. Henceforth, I observed that Demotic is a form of hieroglyphic abbreviation.
(d) From (3) to (1): I entertain the information that last year most of the blackboards and computers in the classrooms of my faculty were changed and upgraded. What I was not aware of is that some of the new blackboards that were installed are interactive. A few days ago, in class, I was pointing at the blackboard and accidentally touched it, immediately a menu with drawing options appeared and so I observed that there is an interactive blackboard in the classroom. This constitutes a purely evidential observation, since it stems from a sensory experience, which provides me with the explicit knowledge of there being an interactive blackboard.
(e) From (2) to (1): I know explicitly that there is a student in my class who is very good at computer programming and I also know that last semester some student from this class made a computer program for inserting boolean operators into a Word-document. But I had not inferred that it was my student who did this. During a class he mentioned his program and showed it to me, thereby I observed that my student has made a computer program. This act coincides with the act of performing a deductive inference, since the action of observation adds the observed information both to awareness-that and to awareness-of, and in this example I was already aware of the information but not yet aware-that it was the case.

The act of receiving external information was first formally discussed in Gerbrandy and Groeneveld (1997) and Plaza (1989). Afterwards the 'Public Announcement Logic' incorporated this action in a multi-agent setting, which has been further developed in different ways. What is worth highlighting, from an epistemological point of view, is that this action represents the way in which external information can be part of the agent's information and that it corresponds automatically to a justified type of information.

5.2.6 Conclusive Remarks on Epistemic Actions

The *combination* of the five epistemic actions I here set forth can serve for representing most of the situations that happened to a single agent[4] regarding her informational attitudes. The knowledge-forming process can be explained by this actions as well as the related processes of knowledge-loss. The establishment of both awareness-of and awareness-that as central parts of the information of the agent and the definition of the different actions form the intended correlation to the epistemological view of AJI.

It is also interesting to point out that some actions can have 'side effects', meaning that after an agent, for example, observes a conjunction of two propositions ($\varphi = p \wedge q$) she will not only be aware-that φ is the case and aware-of φ, but also aware of the components of this formula, that is, she will be automatically aware of p and of q. Formally this will be possible, since the awareness-of set will be formed by atomic propositions and closed under subformulas.

Another relevant point to mention, as I explained regarding the actions of *deductive inference* and *observation*, is that these actions constitute the needed *justification* for the explicit aware knowledge. The need of a justification stems from the epistemological point of view and its correlate in the EAK-Schema is the capability of the agent of performing these actions that increase her Explicit Aware Knowledge.

5.3 The EAK-Schema as a Conceptual Bridge

The presented conceptual schema has served for concluding both thematic threads presented at the beginning of this chapter (Sect. 5.1). I will now emphasise the main features of the EAK-Schema that have achieved this task.

From Epistemology to Logic. Taking Epistemology as a starting point of this research, what has been achieved is a theoretical framework for a logical system (or formal model) inspired by the basic concepts from one specific epistemological view, namely, Awareness Justification Internalism (AJI). This view states that knowledge is justified true belief (JTB-view) and that the Awareness Requirement needs to be fulfilled, that is, the agent needs to be aware of the grounds of whatever justifies her belief. The concept of *belief* has been reinterpreted as 'information', adopting a more general understanding, permitting thus more informational attitudes of the

[4]This approach could be extended to a multi-agent scenario.

agent. The *true information* is achieved with the concept of *awareness-that*, alluding to the acknowledgement of the truth of a given information by the agent. The *Awareness Requirement* is represented with the notion of *awareness-of*, understood as the collection of information the agent has at her disposal.

Finally, the concept of *justification* is given by the *epistemc actions* that stem from the dynamic point of view. The information the agent has at her disposal (the purple ellipse in Fig. 5.6 representing awareness-of) and the true information the agent has acknowledged (the orange ellipse in Fig. 5.6 representing awareness-that) can be changed due to the different actions that have been presented. The central notion of *Explicit Aware Knowledge* represents the real knowledge that AJI is defining and the different actions show distinct strategies the agent can follow for obtaining this 'real' knowledge. As such, the justification is given by the compilation of the different epistemic actions that have turned the information into Explicit Aware Knowledge.

In short, the EAK-Schema (the diagram) representing the informational attitudes of an epistemic agent and the different actions that can move pieces of information from one zone to another in the diagram serves as a theoretical foundation for a logical system that applies the main notions from AJI.

From Logic to Philosophy of Logic. The other direction of the completed research stems from an analysis of the main concepts that underlie contemporary logical systems representing agents with limited resources and including a notion of explicit knowledge. By unravelling these concepts and considering them from their appropriateness for representing the information of human beings (the epistemological perspective) instead of abstract 'epistemic agents', I have selected those concepts or strategies that fitted best the intended correlation.

The EAK-Schema incorporates all these concepts and, in this manner, can be considered a re-interpretation of the logical concepts from a philosophical point of view, i.e. Philosophy of Logic. These notions are the following ones:

- Explicit knowledge is prior in the system: the formation of the diagram and its definitions start with *awareness-that*, which combined with *awareness-of* gives rise to the Explicit Aware Knowledge.
- Implicit knowledge is what can be deduced from the explicit one: the closure under logical consequence of the collection of aware-that information (the green ellipse in Fig. 5.6) represents every information the agent can obtain.
- The concept of 'awareness-of' determines what the agent is actually entertaining: the purple ellipse in the diagram on Fig. 5.6 representing awareness-of and its overlapping with the previous two ellipses represents the need for the true information to be aware-of in order to result in Explicit Aware Knowledge.
- Awareness-that has to be considered semantically: since this concept is already a type of knowledge, in the sense that the agent acknowledges the truth of the information, the formal treatment of this notion will have to be semantic, as 'opposed' to the syntactic understanding.
- Awareness-of, by contrast, has to be considered syntactically: the information contained in the awareness-of corresponds to every type of information, as such, it

has to be considered as a list of information that is defined for the agent at a given moment.

- The epistemic actions will change either the awareness-of or the awareness-that or both: the components of the formal model that will be altered due to an epistemic action have to be awareness-of and awareness-that, that is, the list of information the agent has at her disposal and the function that determines the truth of her aware-that information, respectively.

Some advantages. The EAK-Schema, in its two strands, allows for a finer representation of the informational attitudes of the agent, while it also serves as a logical correlate to the epistemological view of AJI. Regarding the *informational attitudes*, with the EAK-Schema I am able to represent not only what the agent really knows (Explicit Aware Knowledge), but also what she has known at a given moment but is not considering at this stage (aware-that information she is not aware-of). I can as well represent what she does not know yet (implicit knowledge), but will get to know after she deduces it from her EAK and likewise the information she will only obtain after she first becomes aware of it and then deduces it (implicit awareness-that). The EAK-Schema also shows the information whose truth the agent cannot reach yet, but is currently entertaining (awareness-of), and the information she cannot reach by any means (the whole domain). Being the Explicit Aware Knowledge the only form of 'real knowledge' the agent has, the remaining forms of information represent several types of *obtainable knowledge* the agent might reach by different *sequences of actions*.

With respect to the EAK-Schema as a *correlation* for the view of Awareness Justification Internalism, I have constructed a conceptual structure formed by logical concepts inspired in AJI. This means that now there is a direct link between a logical structure (and the corresponding formal model to be presented in Chap. 6) and a purely theoretical epistemological view. Since one of the main motivations for engaging in this research has been the current disconnection between Epistemic Logic and Epistemology, I can now conclude that I have established a bridge that, though still small, highlights the common notions and relations between both fields.

The bridge I have built can be seen as a first step towards the approximation of Epistemology and Epistemic Logic. It is not *the* first step that has been done in this direction, but I consider it a relevant one. Nevertheless, the fact that I chose one specific epistemological view (AJI) inevitably limits the possible logical outcomes, implying, by contrast, a much richer overview of the informational attitudes at hand. My purpose has been to select a very concrete and defined epistemological view and on the basis of it, create the notional bridge to the few involved logical concepts. With only two main notions, namely, *awareness-of* and *awareness-that*, and their different combinations, I have constructed the EAK-Schema that represents the human subject that AJI defines.

References

Ågotnes, T., & Alechina, N. (2007). The dynamics of syntactic knowledge. *Journal of Logic and Computation*, *17*(1), 83–116.

Duc, H. N. (1997). Reasoning about rational, but not logically omniscient agents. *Journal of Logic and Computation*, *7*(5), 633–648.

Fagin, R., & Halpern, J. Y. (1988). Belief, awareness, and limited reasoning. *Artificial Intelligence*, *34*(1), 39–76.

Fernández-Duque, D., Nepomuceno-Fernández, Á., Sarrión-Morillo, E., Soler-Toscano, F., & Velázquez-Quesada, F. R. (2015). Forgetting complex propositions. *Logic Journal of the IGPL*, *23*(6), 942–965.

Fernández-Fernández, C., & Velázquez-Quesada, F. R. (2018). Reconsidering the 'Ingredients' of Explicit Knowledge. In P. Arazim & T. Lávička (Eds.), *The Logica Yearbook 2017* (pp. 47–60). London: College Publications.

Gerbrandy, J., & Groeneveld, W. (1997). Reasoning about information change. *Journal of Logic, Language and Information*, *6*(2), 147–169.

Gettier, E. (1963). Is justified true belief knowledge? *Analysis*, *23*(6), 121–123.

Hill, B. (2010). Awareness dynamics. *Journal of Philosophical Logic*, *39*(2), 113–137.

Jago, M. (2009). Epistemic logic for rule-based agents. *Journal of Logic, Language and Information*, *18*(1), 131–158.

Konolige, K. (1984). *A Deduction Model of Belief and its Logics*. Ph.D. thesis, Computer Science Department, Stanford University.

Plaza, J. A. (1989). Logics of public communications. In M. L. Emrich, M. S. Pfeifer, M. Hadzikadic, & Z. W. Ras (Eds.), *Proceedings of the 4th International Symposium on Methodologies for Intelligent Systems* (pp. 201–216). Tennessee: ORNL.

van Benthem, J., & Velázquez-Quesada, F. R. (2010). The dynamics of awareness. *Synthese*, *177*(Suppl.-1), 5–27.

van Ditmarsch, H., & French, T. (2011). Becoming aware of propositional variables. In M. Banerjee & A. Seth (Eds.), *Logic and Its Applications—4th Indian Conference, ICLA 2011, Delhi, India, January 5–11, 2011. Proceedings*. Volume 6521 of Lecture Notes in Computer Science (pp. 204–218). Berlin: Springer.

van Ditmarsch, H., French, T., & Velázquez-Quesada, F. R. (2012). Action models for knowledge and awareness. In W. van der Hoek, L. Padgham, V. Conitzer, & M. Winikoff (Eds.), *AAMAS 2012, Valencia, Spain, June 4–8, 2012* (pp. 1091–1098). Richland: IFAAMAS.

van Ditmarsch, H., Herzig, A., Lang, J., & Marquis, P. (2009). Introspective forgetting. *Synthese*, *169*(2), 405–423.

Velázquez-Quesada, F. R. (2013). Explicit and implicit knowledge in neighbourhood models. In D. Grossi, O. Roy, & H. Huang, (Eds.), *Logic, Rationality, and Interaction—4th International Workshop, LORI 2013, Hangzhou, China, October 9–12, 2013, Proceedings*. Volume 8196 of Lecture Notes in Computer Science (pp. 239–252). Berlin: Springer.

Chapter 6
Formal Model for Explicit Aware Knowledge

Abstract This chapter presents a formal model that constitutes the logical correlation to the conceptual schema for Explicit Aware Knowledge (EAK-Schema). Using neighbourhood semantics this model depicts the informational attitudes of a non-omniscient epistemic agent. It represents the *awareness of* and *awareness that* of the agent and the different combinations of them that result in Explicit Aware Knowledge, implicit knowledge and other forms of knowledge. The model also includes the formalisation of epistemic actions and discusses the main properties of the informational attitudes and the corresponding epistemic actions that change these attitudes.

Keywords Formal model · Awareness · Neighbourhood semantics · Augmentation operation · Epistemic actions

The EAK-Schema is comprised of several theoretical logical concepts that are suitable of constituting a formal model. There are different strategies for building this model, what follows is one possible approximation. Having its roots in the logical notions from the EAK-Schema, I will now present a formal model inspired by the work in Velázquez-Quesada (2013), where neighbourhood models are used for representing explicit and implicit knowledge (as analysed in Sect. 3.2). It is important to recall that this is merely one possible formal representation of the main epistemic concepts this research has proposed and that it does not constitute a logical 'system' yet, since the axiomatisation has not been developed.[1]

I will first introduce and recall some of the epistemic concepts that have been provided in this research, then present the basic framework (the *Awareness Neighbourhood Model* (ANM)) and afterwards analyse some properties and relationships between the formalised epistemic concepts. Next, I will turn my attention to the formalisation of the epistemic actions, analysing also some basic properties with respect to Explicit Aware Knowledge (EAK).

---

[1]Most of what is stated in this chapter has been published in Fernández-Fernández and Velázquez-Quesada (2020). A previous work, containing a first version of this model, is published in Fernández-Fernández and Velázquez-Quesada (2019).

C. Fernández-Fernández, *Awareness in Logic and Epistemology*,
Logic, Epistemology, and the Unity of Science 52,
https://doi.org/10.1007/978-3-030-69606-1_6

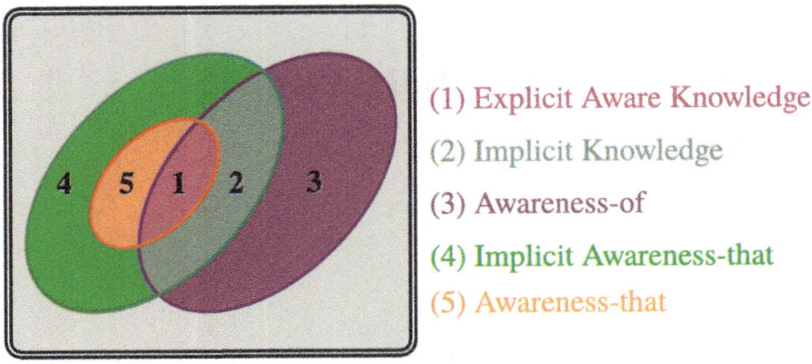

Fig. 6.1 EAK-Schema diagram

The central idea this formal model will reproduce is the distinction between the concepts of *awareness-of* and *awareness-that*, separating thus the mere fact of entertaining some information (being *aware of* φ; a matter of attention) from acknowledging that the information is indeed the case (being *aware that* φ holds; having some sort of 'evidence' for it). The combination of these concepts results in the different epistemic notions (or informational attitudes) this model will be able to ascribe to a single agent.

The following diagram for the EAK-Schema (Fig. 6.1) shows the five main epistemic concepts that represent the ways in which the agent can 'sustain' a given information at a certain moment. Recall that the central notion this diagram hints towards is Explicit Aware Knowledge (EAK), meaning, a true information the agent has acknowledged as such and is entertaining at the moment of evaluation, i.e. information she is both aware-that and aware-of, respectively.

Apart from the 'pure' awareness-of (the purple ellipse in the diagram), the 'pure' awareness-that (the orange ellipse in the diagram) and its overlapping into EAK (1) (the central region in the diagram), this figure also depicts what can be deduced from the awareness-that, that is, two forms of implicit information (the green ellipse circling the orange one). On the one hand, there is *implicit knowledge* (2), referring to what the agent is entertaining, has not recognised as true, but will do after some deductive reasoning (represented by the overlap of awareness-of and the implicit awareness-that); and on the other hand, there is *implicit awareness-that* (4), that is, information the agent could deduce from her awareness-that, once she becomes aware-of it.

This formal model will be able two reproduce three central ideas that derive from the EAK-Schema and that, to the best of my knowledge, have not been formalised in this manner yet (meaning, using awareness-of and awareness-that as central notions in a neighbourhood structure). First, a 'real' agent might not entertain all relevant possibilities (she might lack awareness-of), and she might not acknowledge as true the logical consequences of what she has already accepted (she might lack awareness-that). Second, considering these two different forms of awareness together allows a richer (and thus more realistic) representation of epistemic states. For example, it is possible to have an agent that, while trying to prove a mathematical theorem, has accepted the hypotheses as true and is aware of the conclusion, and still fails to accept the latter (yet). Thirdly, even though an agent does not need to be aware of all possibilities and does not need to acknowledge all logical consequences of what she already has, this does not mean that she is logically ignorant: she might become aware of new alternatives, and she might perform deductive inferences to find out that some additional facts are indeed the case. In other words, a 'logically competent' agent does not have to be logically omniscient; it is enough for her to be able to perform different epistemic actions to change (and, sometimes, improve) her information.

6.1 Awareness Neighbourhood Model: Main Structure

This section introduces the basic framework for depicting Explicit Aware Knowledge (EAK) as the combination of *awareness-of* and *awareness-that*. On the semantic side, the structure used for modelling such concepts is a neighbourhood model (Scott 1970; Montague 1970; see Chellas 1980, Chap. 7 and Pacuit 2017 for a modern presentation) extended with a global atomic awareness set (cf. Fagin and Halpern 1988). The neighbourhood function is used for representing what the agent is aware-that, with each world's neighbourhood understood as a list containing the semantic representation of the formulas the agent has acknowledged as true. This provides that the notion of awareness-that is not closed under logical consequence, as one gets when using more standard relational possible worlds models. The atomic awareness set is used for representing what the agent is aware-of, which gives a notion of awareness-of that can be understood as the atoms defining the agent's current language (satisfying thus the main features of the EAK-Schema listed in Sect. 5.3).

In this model, let P be a non-empty enumerable set of atomic propositions.

Definition 6.1.1 (*Awareness neighbourhood model*) An *awareness neighbourhood* model (*ANM*) is a tuple $M = \langle W, N, V, A\rangle$ where ***(i)*** W, also denoted as $\mathcal{D}_M$, is a non-empty set (of *possible worlds*); ***(ii)*** $N : W \to \wp(\wp(W))$ is a *neighbourhood function* (assigning a set of sets of worlds to each possible world, with $N(w)$ called the *neighbourhood* of w); ***(iii)*** $V : \mathrm{P} \to \wp(W)$ is an *atomic valuation function* (indicating the set of possible worlds in which each atom is true); ***(iv)*** $A \subseteq \mathrm{P}$ is the *atomic awareness set* (indicating the set of atoms the agent is aware of). ◀

On the syntactic side, the language extends the propositional one with operators for describing the agent's *awareness of* (A^{o}), her *awareness that* (A^{t}), and the deductive closure of the latter ($[*]$).

Definition 6.1.2 (*Language and semantic interpretation*) Formulas φ, ψ of the language $\mathcal{L}$ are given by

$$\varphi, \psi ::= \top \mid p \mid \neg\varphi \mid \varphi \wedge \psi \mid \mathrm{A}^{\mathrm{o}}\varphi \mid \mathrm{A}^{\mathrm{t}}\varphi \mid [*]\varphi$$

with $p \in \mathrm{P}$. Formulas of the form $\mathrm{A}^{\mathrm{o}}\,\varphi$ are read as *"the agent is aware of φ"*, those of the form $\mathrm{A}^{\mathrm{t}}\,\varphi$ are read as *"the agent is aware that φ"*, and $[*]\,\varphi$ expresses that *"after the agent performs every possible deductive inference, φ holds"*. For any formula $\varphi \in \mathcal{L}$, define its set of atoms, $\mathrm{at}(\varphi)$, in the standard way:

$$\begin{array}{lll} \mathrm{at}(\top) := \varnothing, & \mathrm{at}(\neg\varphi) := \mathrm{at}(\varphi), & \mathrm{at}(\mathrm{A}^{\mathrm{o}}\,\varphi) := \mathrm{at}(\varphi), \\ \mathrm{at}(p) := \{p\}, & \mathrm{at}(\varphi \wedge \psi) := \mathrm{at}(\varphi) \cup \mathrm{at}(\psi), & \mathrm{at}(\mathrm{A}^{\mathrm{t}}\,\varphi) := \mathrm{at}(\varphi), \\ & & \mathrm{at}([*]\,\varphi) := \mathrm{at}(\varphi). \end{array}$$

Given an *ANM* $M = \langle W, N, V, A\rangle$ and a formula $\varphi \in \mathcal{L}$, the function $[\![\cdot]\!]^M : \mathcal{L} \to \wp(W)$ returns the set of worlds in M at which φ holds ($[\![\varphi]\!]^M$ is φ's *truth-set*). In its (inductive) definition, the cases for $\top$, atoms and Boolean operators are standard:

$$[\![\top]\!]^M := W, \quad [\![p]\!]^M := V(p), \quad [\![\neg\varphi]\!]^M := W \setminus [\![\varphi]\!]^M, \quad [\![\varphi \wedge \psi]\!]^M := [\![\varphi]\!]^M \cap [\![\psi]\!]^M .$$

For $\mathrm{A}^{\mathrm{o}}\,\varphi$ (awareness of), the formula is either globally true (when the agent is aware of all φ's atoms) or else globally false (otherwise):

$$[\![\mathrm{A}^{\mathrm{o}}\,\varphi]\!]^M := \begin{cases} W \text{ if } \mathrm{at}(\varphi) \subseteq A; \\ \varnothing \text{ otherwise.} \end{cases}$$

For $\mathrm{A}^{\mathrm{t}}\,\varphi$ (awareness that), the formula is true at a world w in M if and only if the truth-set of φ in M is in the neighbourhood of w:

$$[\![\mathrm{A}^{\mathrm{t}}\,\varphi]\!]^M := \left\{ w \in W \mid [\![\varphi]\!]^M \in N(w) \right\}.$$

The remaining modality, $[*]$, is a special case, as its semantic interpretation relies not only on the given M, but also on its *augmentation*: the model M^* that results from making the neighbourhood of each world a set that contains the neighbourhood's *core*, $\bigcap N(w)$, and is closed under supersets.[2] This corresponds to the *dynamic epistemic logic* approach (*DEL*; van Ditmarsch et al. 2008; van Benthem 2011), that defines operations that change the underlying model, and then uses modalities that are interpreted over the resulting models. More precisely, from the given $M = \langle W, N, V, A\rangle$, define $M^* = \langle W, N^*, V, A\rangle$ with N^* such that

$$N^*(w) := \left\{ U \subseteq W \mid \bigcap N(w) \subseteq U \right\}.$$

Then,

$$[\![[*]\,\varphi]\!]^M := [\![\varphi]\!]^{M^*} .$$

As will be recalled (Sect. 6.2.2), in M^* the modality A^{t} behaves as the $\Box$-operator does in relational models; this is the reason behind the intuitive reading of formulas

[2]For more details on this model construction, see Chellas (1980), Sect. 7.3.

of the form $[*]\varphi$, and the reason why here $[*]$ is also called the *deductive closure* modality.

Satisfiability and *validity* are defined in the standard way, with the latter denoted as usual ($\Vdash \varphi$). ◀

Thus, while *awareness-that* (A^t) corresponds to the (local) neighbourhood function N, *awareness-of* (A^o) is generated by the (global) set of atoms A.

6.2 Properties and Relationships Between Epistemic Concepts

The formal framework introduced above allows me to provide formal definitions for the notions the diagram on Fig. 6.1 sketches. Besides *awareness-of* and *awareness-that*, the epistemic concepts that will be discussed from a formal point of view, are the following ones:

- **Explicit Aware Knowledge** (EAK) is what the agent is entertaining (i.e. is aware of) and has also acknowledged as true (i.e. is aware that):

$$K_{Ex}\,\varphi := A^o\,\varphi \wedge A^t\,\varphi.$$

  In this sense, EAK is what the agent 'really' knows at the given stage.
- **Implicit Knowledge** is what the agent is already entertaining (i.e. is aware-of) and will acknowledge as true after applying all possible deductive inferences:

$$K_{Im}\,\varphi := A^o\,\varphi \wedge [*]\,A^t\,\varphi.$$

  In other words, implicit knowledge is what the agent can deduce from what she knows explicitly. As mentioned regarding the EAK-Schema and showed by the above formalisation, this knowledge is logically meaningful, though theoretically it must be considered a 'potential knowledge', since there is no 'real' correlate to it.

Thus, while EAK is what the agent has at the given stage, implicit knowledge is all she will have after applying a full deductive closure to her aware awareness-that.

Explicit Aware and Implicit Knowledge require *awareness of*, as both deal with the possibilities the agent is currently entertaining. But there are similar notions of knowledge, having the same relationship with each other (one the closure under deductive inference of the other), but staying outside of the possibilities the agent is currently pondering. The first one can be called **'disassociated' knowledge** and is what the agent has acknowledged as true but is not currently entertaining (that is, what will become explicitly known after she becomes aware of it), and is given by $K_{Ex}^{-o}\,\varphi := \neg\,A^o\,\varphi \wedge A^t\,\varphi$ (from a theoretical point of view, this 'knowledge' corresponds to the *awareness-that* outside the scope of the awareness-of).

The second one can be named **currently 'unreachable' knowledge**; it is what the agent is not currently entertaining and yet she can deduce from what she has acknowledged as true (i.e. what she could deduce after becoming aware of what she is aware-that), and is given by $\mathrm{K}_{Im}^{-o}\,\varphi := \neg\,\mathrm{A}^{\mathrm{o}}\,\varphi \wedge [*]\,\mathrm{A}^{\mathrm{t}}\,\varphi$ (theoretically this knowledge corresponds to the *implicit awareness-that*).

In what follows I will discuss the properties some of these epistemic concepts obtain under the given semantic model and definitions. The focus will be the two primitive concepts, *awareness-of* and *awareness-that*, as well as *Explicit Aware Knowledge* and *Implicit Knowledge*.

6.2.1 Basic Properties and Relationships

Awareness of This concept, A^{o}, corresponds to what the agent *entertains*. Thus, *awareness-of* is a matter of attention, and by itself it does not imply any attitude pro or con: the agent may have accepted the formula as true, as false, or she might not have any inclination about it.

Semantically, A^{o} is defined in terms of a global set of atomic propositions, the atomic awareness set A. More precisely, given an *ANM* model $M = \langle W, N, V, A\rangle$, the agent is aware of a given formula φ at a world $w \in W$ if and only if all atoms occurring in φ belong to this set, $\mathrm{at}(\varphi) \subseteq A$. As a consequence of this, the agent is aware of the concept of 'truth':

$$\Vdash \mathrm{A}^{\mathrm{o}}\,\top.$$

This does not say that the agent is aware that this 'truth' holds everywhere; it simply states that she entertains such concept. Technically, the reason is that $\top$, a primitive in the language (thus not defined as an abbreviation of the form $p \vee \neg p$, as in other proposals) does not contain any atomic proposition.

Still, despite being aware of the concept of truth, the agent does not need to be aware of formulas that are true in every possible situation (that is, she does not need to be aware of valid formulas):

- $\Vdash \varphi$ does not imply $\Vdash \mathrm{A}^{\mathrm{o}}\,\varphi$.

Technically, this is the case because a validity might have atoms (e.g., $p \vee \neg p$), and no atom is required to be in the atomic awareness set (e.g., p does not need to be in A); hence, there might be validities the agent does not entertain. Along the same lines, *awareness-of* is not closed under logical equivalence, as the given formulas might involve different atoms:

- $\Vdash \varphi \leftrightarrow \psi$ does not imply $\Vdash \mathrm{A}^{\mathrm{o}}\,\varphi \leftrightarrow \mathrm{A}^{\mathrm{o}}\,\psi$.

Note that *awareness-of* is defined not in terms of a set of formulas (as in the logic of general awareness of Fagin and Halpern 1988) but rather in terms of a set of *atomic propositions*. As discussed already in the mentioned paper, this makes the concept of *awareness-of* closed not only under subformulas but also under superformulas. More precisely,

$$\Vdash \mathrm{A}^{\mathrm{o}}\, \neg\varphi \leftrightarrow \mathrm{A}^{\mathrm{o}}\, \varphi, \qquad \Vdash \mathrm{A}^{\mathrm{o}}\, \mathrm{A}^{\mathrm{o}}\, \varphi \leftrightarrow \mathrm{A}^{\mathrm{o}}\, \varphi,$$

$$\Vdash \mathrm{A}^{\mathrm{o}}(\varphi \wedge \psi) \leftrightarrow (\mathrm{A}^{\mathrm{o}}\, \varphi \wedge \mathrm{A}^{\mathrm{o}}\, \psi), \qquad \Vdash \mathrm{A}^{\mathrm{o}}\, \mathrm{A}^{\mathrm{t}}\, \varphi \leftrightarrow \mathrm{A}^{\mathrm{o}}\, \varphi,$$

$$\Vdash \mathrm{A}^{\mathrm{o}}\, [*]\, \varphi \leftrightarrow \mathrm{A}^{\mathrm{o}}\, \varphi.$$

In particular, note what the formulas on the right column indicate. The first states that awareness of awareness of a formula is equivalent to awareness of the formula. The other two indicate, with some paraphrasing (and read from right to left), that entertaining φ is equivalent to entertaining the possibility of having accepted φ as true (the second) and also equivalent to entertaining the possibility of obtaining φ after deductive inference (the third).

An important difference between the *awareness-of* discussed here and the original one in Fagin and Halpern (1988) is that, while the latter is local (what belongs to the awareness set assigned to the given evaluation point), the one proposed here is global (what belongs to the atomic awareness set assigned to the whole model). An intermediate position is held in the relational-model-based approach of Grossi and Velázquez-Quesada (2015), where (atom-based) *awareness of* is defined as those formulas whose atoms appear in the awareness set of all worlds the agent cannot distinguish from the evaluation point. The definition makes sense in the multi-agent setting that the referred paper studies; in the single-agent case examined here, one can assume that the worlds in the model are exactly those that are relevant for the agent under discussion, thus making the two definitions (that of Grossi and Velázquez-Quesada 2015 and the one proposed here) conceptually equivalent.

Awareness that. The concept of *awareness-that*, A^{t}, understood as what the agent has accepted/acknowledged as true, is a form of 'explicit' information (as discussed in detail in Chap. 2). Even though *awareness-that* is acknowledgement of truth, such acceptance does not imply by itself that the agent is entertaining such piece of information (she might have accepted it before, and then moved on to a different topic), and therefore it does not imply explicit knowledge either.

Semantically, A^{t} is defined as what appears in the neighbourhood of the current evaluation point.[3] In fact, the neighbourhood of a given world, a set of sets of worlds, can be understood as the list of formulas the agent has acknowledged as true, the important point being that these formulas are not represented syntactically (as a string of symbols), but rather semantically (as the set of those worlds in the model in which the formula holds). Because of this purely semantic representation, the concept of awareness-that has an important closure property: it is *closed under logical equivalence*:

$$\Vdash \varphi \leftrightarrow \psi \quad \text{implies} \quad \Vdash \mathrm{A}^{\mathrm{t}}\, \varphi \leftrightarrow \mathrm{A}^{\mathrm{t}}\, \psi.$$

[3]Note that A^{t} uses the 'set' semantic interpretation of the $\Box$-operator in neighbourhood models. (Recall: the 'subset' semantic interpretation makes $\Box\varphi$ true at w in M not only when $[\![\varphi]\!]^M$ is in $N(w)$, but also when any of its *subsets* is: $[\![\Box\varphi]\!]^M := \{w \in W \mid \text{there is } U \in N(w) \text{ such that } U \subseteq [\![\varphi]\!]^M\}$.)

Thus, the agent has indeed some omniscience: at the level of acknowledgement, she cannot tell apart formulas that are true in exactly the same worlds in all models.[4] Still, this does not mean that the agent's EAK is closed under logical equivalence: the concept of awareness-of will help to distinguish between logically equivalent formulas.

Closure under logical equivalence is the only closure property the notion of awareness-that has. Different from other semantic representations of information, as the $\Box$-operator in standard epistemic logic under relational models, ***(i)*** validities do not need to be acknowledged as true, ***(ii)*** awareness-that is not closed under conjunction introduction, and ***(iii)*** neither under conjunction elimination.

(i) $\Vdash \varphi$ does not imply $\Vdash \mathrm{A}^{\mathrm{t}}\,\varphi$,
(ii) $\nVdash (\mathrm{A}^{\mathrm{t}}\,\varphi \wedge \mathrm{A}^{\mathrm{t}}\,\psi) \rightarrow \mathrm{A}^{\mathrm{t}}(\varphi \wedge \psi)$,
(iii) $\nVdash \mathrm{A}^{\mathrm{t}}(\varphi \wedge \psi) \rightarrow \mathrm{A}^{\mathrm{t}}\,\varphi$ and $\nVdash \mathrm{A}^{\mathrm{t}}(\varphi \wedge \psi) \rightarrow \mathrm{A}^{\mathrm{t}}\,\psi$.

The reason for the failure of these properties is that no neighbourhood needs to have any closure property. In particular, ***(i)*** none of them needs to contain the whole domain (so $\mathcal{D}_M$, the truth-set of any validity, does not need to be in $N(w)$), ***(ii)*** none of them needs to be closed under intersections (so $[\![\varphi]\!]^M, [\![\psi]\!]^M \in N(w)$ does not imply $[\![\varphi]\!]^M \cap [\![\psi]\!]^M = [\![\varphi \wedge \psi]\!]^M \in N(w)$), and ***(iii)*** none of them needs to be closed under supersets (so $[\![\varphi \wedge \psi]\!]^M = [\![\varphi]\!]^M \cap [\![\psi]\!]^M \in N(w)$ implies neither $[\![\varphi]\!]^M \in N(w)$ nor $[\![\psi]\!]^M \in N(w)$). Because of the failure of conjunction introduction/elimination, awareness-that is not closed under modus ponens[5]:

- $\nVdash \mathrm{A}^{\mathrm{t}}(\varphi \rightarrow \psi) \rightarrow (\mathrm{A}^{\mathrm{t}}\,\varphi \rightarrow \mathrm{A}^{\mathrm{t}}\,\psi)$.

Hence, awareness-that is *not closed under logical consequence*.

Awareness-of and Awareness-that. A property relating A^{o} and A^{t} has already been discussed ($\Vdash \mathrm{A}^{\mathrm{o}}\,\mathrm{A}^{\mathrm{t}}\,\varphi \leftrightarrow \mathrm{A}^{\mathrm{o}}\,\varphi$). Yet, attending standard Awareness Logic, the fact that awareness-of is global might suggest a further relationship: that the agent 'knows her own awareness'. Indeed, in the original (Fagin and Halpern 1988), the fact that the awareness set of all worlds in the model is the same implies not only $\Vdash \mathrm{A}\,\varphi \rightarrow \Box\,\mathrm{A}\,\varphi$ (if the agent is aware of φ, then she knows this [implicitly]) but also $\Vdash \neg\,\mathrm{A}\,\varphi \rightarrow \Box\,\neg\,\mathrm{A}\,\varphi$ (if she is not aware of φ, then she knows this [implicitly]).

In the present model, analogous properties do not need to hold. The agent does not need to acknowledge any formula, even if it holds in every world of the model; thus, she does not need to acknowledge her *awareness-of* and neither her *unawareness-of*.

- $\nVdash \mathrm{A}^{\mathrm{o}}\,\varphi \rightarrow \mathrm{A}^{\mathrm{t}}\,\mathrm{A}^{\mathrm{o}}\,\varphi$,
- $\nVdash \neg\,\mathrm{A}^{\mathrm{o}}\,\varphi \rightarrow \mathrm{A}^{\mathrm{t}}\,\neg\,\mathrm{A}^{\mathrm{o}}\,\varphi$.

---

[4] When awareness-that is understood as explicit knowledge it does not have this closure property (like in Konolige 1984; Grossi and Velázquez-Quesada 2015, for example), simply because it is represented by a set of formulas that is not required to have any closure property.

[5] In a semantic setting, a modus ponens can be understood as a three-step process: conjunction introduction (from $[\![\varphi]\!]^M$ and $[\![\varphi \rightarrow \psi]\!]^M$ to $[\![\varphi \wedge (\varphi \rightarrow \psi)]\!]^M$), logical equivalence (from the latter to $[\![\varphi \wedge \psi]\!]^M$) and conjunction elimination (from the now-latter to $[\![\psi]\!]^M$).

Explicit Aware Knowledge. This concept, K_{Ex}, is defined as those pieces of information the agent has acknowledged as true and is currently entertaining, i.e. the agent is both aware-of and aware-that: $\mathrm{K}_{Ex}\,\varphi := \mathrm{A}^{\mathrm{o}}\,\varphi \wedge \mathrm{A}^{\mathrm{t}}\,\varphi$.

About explicit aware knowledge of validities. Since the agent needs to be neither aware of them nor aware that they are the case, she does not need to know explicitly any validity. Moreover: none of the awareness concepts is, by itself, enough to guarantee that a validity is explicitly known:

- $\Vdash \varphi$ implies neither $\Vdash \mathrm{K}_{Ex}\,\varphi$ nor $\Vdash \mathrm{A}^{\mathrm{o}}\,\varphi \to \mathrm{K}_{Ex}\,\varphi$ nor $\Vdash \mathrm{A}^{\mathrm{t}}\,\varphi \to \mathrm{K}_{Ex}\,\varphi$.

About closure under logical equivalence. Even though awareness-that has this property, Explicit Aware Knowledge does not need to, as the agent might be aware of a formula without being aware of a logically equivalent one. Thus,

- $\Vdash \varphi \leftrightarrow \psi$ does not imply $\Vdash \mathrm{K}_{Ex}\,\varphi \leftrightarrow \mathrm{K}_{Ex}\,\psi$.

Still, awareness-of is the only piece that is missing. Thus, in particular, if two formulas are logically equivalent and the agent is aware of the second, then explicit knowledge of the first implies explicit knowledge of the second:

$$\Vdash \varphi \leftrightarrow \psi \quad \text{implies} \quad \Vdash \mathrm{A}^{\mathrm{o}}\,\psi \to (\mathrm{K}_{Ex}\,\varphi \to \mathrm{K}_{Ex}\,\psi).$$

About closure under modus ponens. Since awareness-that lacks this property, EAK lacks it too:

- $\nVdash \mathrm{K}_{Ex}(\varphi \to \psi) \to (\mathrm{K}_{Ex}\,\varphi \to \mathrm{K}_{Ex}\,\psi)$.

The lack of this property is already shared by the explicit knowledge of Fagin and Halpern (1988) (defined, recall, as implicit knowledge, $\Box\,\varphi$, plus awareness-of, $\mathrm{A}\,\varphi$). However, different from the mentioned proposal, here being aware of the consequent does not give the agent explicit knowledge about it:

- $\nVdash \mathrm{K}_{Ex}(\varphi \to \psi) \to \big((\mathrm{K}_{Ex}\,\varphi \wedge \mathrm{A}^{\mathrm{o}}\,\psi) \to \mathrm{K}_{Ex}\,\psi\big)$.

In other words, the agent might know explicitly an implication and its antecedent, and she might even entertain the consequent, and still she does not need to know explicitly that this consequent is indeed the case. In fact, assuming the agent entertains the consequent does not provide anything new. By explicitly knowing the implication, the agent is aware of it, and thus she is aware of its consequent, as awareness-of is closed under subformulas. What the agent misses is the acknowledgement that the implication's consequence is indeed the case[6]:

$$\Vdash \mathrm{K}_{Ex}(\varphi \to \psi) \to \big((\mathrm{K}_{Ex}\,\varphi \wedge \mathrm{A}^{\mathrm{t}}\,\psi) \to \mathrm{K}_{Ex}\,\psi\big).$$

Crucially, acknowledgement can be reached via some *epistemic action*.

[6] In this, the present setting coincides with the one in Grossi and Velázquez-Quesada (2015).

6.2.2 *Effects of the Augmentation Operation*

The augmentation operation makes the neighbourhood of each world a set that is closed under supersets and contains the neighbourhood's core ($\bigcap N(w)$). It is well-known (e.g., Chellas 1980, Theorem 7.9) that in the resulting model, the *augmented* model M^*, the operator A^{t} behaves as the standard $\Box$-operator does in relational models. Hence, occurrences of A^{t} under the scope of the modality $[*]$ can be understood as what the agent will acknowledge as true after she applies every possible deductive inference.[7] In this sense, the augmentation operation makes the agent's awareness-that closed under logical consequence, and hence it can be understood as a *full deductive inference* operation over A^{t}.

Awareness-that after full deductive inference. As a consequence of the augmentation operation, the agent acknowledges every validity as true:

$$\Vdash \varphi \quad \text{implies} \quad \Vdash [*]\,\mathrm{A}^{\mathrm{t}}\,\varphi.$$

Moreover, her awareness-that becomes closed under conjunction introduction,

$$\Vdash [*]\,\big((\mathrm{A}^{\mathrm{t}}\,\varphi \wedge \mathrm{A}^{\mathrm{t}}\,\psi) \rightarrow \mathrm{A}^{\mathrm{t}}(\varphi \wedge \psi)\big),$$

and also under conjunction elimination,

$$\Vdash [*]\,\big(\mathrm{A}^{\mathrm{t}}(\varphi \wedge \psi) \rightarrow \mathrm{A}^{\mathrm{t}}\,\varphi\big) \qquad \text{and} \qquad \Vdash [*]\,\big(\mathrm{A}^{\mathrm{t}}(\varphi \wedge \psi) \rightarrow \mathrm{A}^{\mathrm{t}}\,\psi\big).$$

In fact, since the augmentation operation is a total function over *ANM*s (defined for every *ANM*, and producing a single result), the last two properties can be described in a more useful way. For the first, if after the operation the agent acknowledges φ and after the operation she acknowledges ψ, then after the operation she acknowledges both and hence she acknowledges their conjunction:

$$\Vdash ([*]\,\mathrm{A}^{\mathrm{t}}\,\varphi \wedge [*]\,\mathrm{A}^{\mathrm{t}}\,\psi) \rightarrow [*]\,\mathrm{A}^{\mathrm{t}}(\varphi \wedge \psi).$$

For the second, if after the operation the agent acknowledges a conjunction, then after the operation she acknowledges any of its conjuncts:

$$\Vdash [*]\,\mathrm{A}^{\mathrm{t}}(\varphi \wedge \psi) \rightarrow [*]\,\mathrm{A}^{\mathrm{t}}\,\varphi \qquad \text{and} \qquad \Vdash [*]\,\mathrm{A}^{\mathrm{t}}(\varphi \wedge \psi) \rightarrow [*]\,\mathrm{A}^{\mathrm{t}}\,\psi.$$

These two closure properties and the discussed closure under logical equivalence makes A^{t} closed under modus ponens *after* the augmentation operation,

[7] Velázquez-Quesada (2013) (see also Balbiani et al. 2018) already uses this known relationship between neighbourhood models and relational models. Still, the technical details are slightly different, as the referred paper works on the finite-domain case and does not take the concept of awareness-of into account.

$$\Vdash [*]\big(A^t(\varphi \to \psi) \to (A^t\varphi \to A^t\psi)\big),$$

and thus, since the operation is a total function,

$$\Vdash [*]\,A^t(\varphi \to \psi) \to ([*]\,A^t\varphi \to [*]\,A^t\psi).$$

Finally, the operation also affects what the agent has acknowledged about her own awareness-of. Recall the technical reason why the agent did not acknowledge what she entertains: she might not have acknowledged all formulas that are true in all worlds. But this 'closure under deductive inference' operation makes her realise that all globally true formulas are indeed the case. Hence,

$$\Vdash [*](A^o\varphi \to A^t A^o\varphi) \quad \text{and} \quad \Vdash [*](\neg A^o\varphi \to A^t \neg A^o\varphi)$$

or, since the operation does not affect awareness sets,

$$\Vdash A^o\varphi \to [*]\,A^t A^o\varphi \quad \text{and} \quad \Vdash \neg A^o\varphi \to [*]\,A^t \neg A^o\varphi.$$

Implicit knowledge. Knowing the properties of awareness-that after the augmentation operation, it is now time to discuss the properties of the notion of *implicit knowledge*, defined as what the agent currently entertains and will recognize as true *after* performing all possible deductive inferences, $K_{Im}\,\varphi := A^o\varphi \land [*]\,A^t\varphi$.

First, the agent's implicit knowledge does not need to contain every validity,

- $\Vdash \varphi$ does not imply $\Vdash K_{Im}\,\varphi$,

the reason being that she might be unaware of some involved atoms.[8] Nevertheless, different from the EAK case, awareness-of is the only missing piece: the agent knows *implicitly* any validity she is currently entertaining:

$$\Vdash \varphi \quad \text{implies} \quad \Vdash A^o\varphi \to K_{Im}\,\varphi.$$

Similarly, the agent's implicit knowledge is not closed under logical equivalence,

- $\Vdash \varphi \leftrightarrow \psi$ does not imply $\Vdash K_{Im}\,\varphi \leftrightarrow K_{Im}\,\psi$,

the only reason being that the awareness-of requirement might fail. Thus,

$$\Vdash \varphi \leftrightarrow \psi \quad \text{implies} \quad \Vdash (K_{Im}\,\varphi \land A^o\psi) \to K_{Im}\,\psi.$$

Finally, implicit knowledge is closed under both conjunction introduction and conjunction elimination. For the first, ***(i)*** after the augmentation, awareness-that has such property (I use here the alternative version of this), and ***(ii)*** awareness-of is closed under superformulas:

---

[8] This differs from the same notion in Fagin and Halpern (1988), where the agent knows implicitly every validity.

$$\Vdash (\mathrm{K}_{Im}\,\varphi \wedge \mathrm{K}_{Im}\,\psi) \rightarrow \mathrm{K}_{Im}(\varphi \wedge \psi).$$

Analogously, for the second: ***(i)*** after the operation, awareness-that has the property (I use its alternative version), and ***(ii)*** *awareness of* is closed under subformulas:

$$\Vdash \mathrm{K}_{Im}(\varphi \wedge \psi) \rightarrow \mathrm{K}_{Im}\,\varphi \quad \text{and} \quad \Vdash \mathrm{K}_{Im}(\varphi \wedge \psi) \rightarrow \mathrm{K}_{Im}\,\psi.$$

The last three validity-statements, together with awareness-of's closure under subformulas, show that implicit knowledge is closed under *Modus Ponens*:

$$\Vdash \mathrm{K}_{Im}(\varphi \rightarrow \psi) \rightarrow (\mathrm{K}_{Im}\,\varphi \rightarrow \mathrm{K}_{Im}\,\psi).$$

Observe how, while the implicit knowledge of (Fagin and Halpern 1988) contains all validities and is closed under modus ponens (i.e. is closed under logical consequence), implicit knowledge as defined here might not contain all validities and, despite being closed under modus ponens, it does not need to be closed under logical equivalence. More precisely, while implicit knowledge in Fagin and Halpern (1988) is the agent's 'semantic' information (given by the modal operator $\Box$), here it is the closure under modus ponens of what the agent has acknowledged as true ($[*]\,\mathrm{A}^{\mathrm{t}}\,\varphi$) and is currently entertaining ($\mathrm{A}^{\mathrm{o}}\,\varphi$); it is the closure under modus ponens of what the agent knows explicitly.

This point highlights the crucial difference between the understanding of implicit knowledge in this model and in Fagin and Halpern (1988). In the latter, what is needed for implicit knowledge to be explicit is for the agent to be aware of the given formula. However, in the present proposal, knowing a formula implicitly already implies the agent is aware of it. Here, what is needed for implicit knowledge to become explicit aware is then not an act of awareness raising; what is needed is an act of deductive inference.

Moorean phenomena. In proposals dealing with implicit and explicit knowledge, a particular property is recurrent: explicit knowledge is also implicit. In Fagin and Halpern (1988), this follows from the fact that explicit knowledge is defined as the implicit knowledge that satisfies an additional requirement (awareness-of); in settings distinguishing explicit information from implicit one by means of deductive reasoning (e.g., Konolige 1986), this follows from the fact that deductive inference is monotone.

This property, seemingly not only natural but rather essential, fails here:

- $\nVdash \mathrm{K}_{Ex}\,\varphi \rightarrow \mathrm{K}_{Im}\,\varphi.$

The reason being (see Velázquez-Quesada 2013) that what the agent has acknowledged as true at some stage does not need to be acknowledged as true after the augmentation operation, i.e.

Fact 6.2.1 $\nVdash \mathrm{A}^{\mathrm{t}}\,\varphi \rightarrow [*]\,\mathrm{A}^{\mathrm{t}}\,\varphi$.

Proof Take the set of atomic propositions $\{p, q\}$ and the formula $\varphi := \neg\,\mathrm{A}^{\mathrm{t}}\,q$, and consider a four-worlds model $M = \langle W = \{w_1, w_2, w_3, w_4\}, N, V, \varnothing\rangle$ with an empty atomic awareness set, valuation given by $V(p) = \{w_1, w_2\}$ and $V(q) = \{w_1, w_3\}$, and neighbourhood function such that

$$N(w_1) := \{\{w_1, w_2\}, \{w_1, w_3, w_4\}, W\}, \qquad N(w_2) = N(w_3) = N(w_4) := \varnothing.$$

From $[\![\mathrm{A}^{\mathrm{t}}\,q]\!]^M = \varnothing$ (no neighbourhood contains $[\![q]\!]^M = \{w_1, w_3\}$) it follows that $[\![\neg\,\mathrm{A}^{\mathrm{t}}\,q]\!]^M = W$; hence, $[\![\neg\,\mathrm{A}^{\mathrm{t}}\,q]\!]^M \in N(w_1)$ and therefore $w_1 \in [\![\mathrm{A}^{\mathrm{t}}\,\neg\,\mathrm{A}^{\mathrm{t}}\,q]\!]^M$. However, observe the neighbourhood function of the augmented model M^*:

$$N^*(w_1) = \{U \subseteq W \mid w_1 \in U\}, \qquad N^*(w_2) = N^*(w_3) = N^*(w_4) = \{W\}.$$

Note how $[\![\mathrm{A}^{\mathrm{t}}\,q]\!]^{M^*} = \{w_1\}$ (only w_1's neighbourhood includes $[\![q]\!]^{M^*} = \{w_1, w_3\}$), so $[\![\neg\,\mathrm{A}^{\mathrm{t}}\,q]\!]^{M^*} = \{w_2, w_3, w_4\}$. Then, $[\![\neg\,\mathrm{A}^{\mathrm{t}}\,q]\!]^{M^*} \notin N^*(w_1)$ so $w_1 \notin [\![\mathrm{A}^{\mathrm{t}}\,\neg\,\mathrm{A}^{\mathrm{t}}\,q]\!]^{M^*}$, that is, $w_1 \notin [\![[*]\,\mathrm{A}^{\mathrm{t}}\,\neg\,\mathrm{A}^{\mathrm{t}}\,q]\!]^M$. □

Thus, while in M the agent has acknowledged $\neg\,\mathrm{A}^{\mathrm{t}}\,q$ as true at w_1 (i.e. $w_1 \in [\![\mathrm{A}^{\mathrm{t}}\,\neg\,\mathrm{A}^{\mathrm{t}}\,q]\!]^M$), the operation changes this: in M^*, the agent has not acknowledged $\neg\,\mathrm{A}^{\mathrm{t}}\,q$ as true at w_1 (i.e. $w_1 \notin [\![[*]\,\mathrm{A}^{\mathrm{t}}\,\neg\,\mathrm{A}^{\mathrm{t}}\,q]\!]^M$). One just needs to make the agent aware of the involved formula $\neg\,\mathrm{A}^{\mathrm{t}}\,q$ (e.g., take $A := \{p, q\}$) to obtain a model (M) and a world (w_1) in which the agent knows a formula ($\neg\,\mathrm{A}^{\mathrm{t}}\,q$) explicitly,

$$\begin{aligned} w_1 \in [\![\mathrm{K}_{Ex}\,\neg\,\mathrm{A}^{\mathrm{t}}\,q]\!]^M &= [\![\mathrm{A}^{\mathrm{o}}\,\neg\,\mathrm{A}^{\mathrm{t}}\,q \;\wedge\; \mathrm{A}^{\mathrm{t}}\,\neg\,\mathrm{A}^{\mathrm{t}}\,q]\!]^M \\ &= [\![\mathrm{A}^{\mathrm{o}}\,\neg\,\mathrm{A}^{\mathrm{t}}\,q]\!]^M \cap [\![\mathrm{A}^{\mathrm{t}}\,\neg\,\mathrm{A}^{\mathrm{t}}\,q]\!]^M, \end{aligned}$$

and yet she does not know it implicitly,

$$\begin{aligned} w_1 \notin [\![\mathrm{K}_{Im}\,\neg\,\mathrm{A}^{\mathrm{t}}\,q]\!]^M &= [\![\mathrm{A}^{\mathrm{o}}\,\neg\,\mathrm{A}^{\mathrm{t}}\,q \;\wedge\; [*]\,\mathrm{A}^{\mathrm{t}}\,\neg\,\mathrm{A}^{\mathrm{t}}\,q]\!]^M \\ &= [\![\mathrm{A}^{\mathrm{o}}\,\neg\,\mathrm{A}^{\mathrm{t}}\,q]\!]^M \cap [\![[*]\,\mathrm{A}^{\mathrm{t}}\,\neg\,\mathrm{A}^{\mathrm{t}}\,q]\!]^M. \end{aligned}$$

So, is there some fundamental problem with the current proposal? To answer this, first note how the 'explicit is implicit' property does not always fail. In fact, it holds for a large class of formulas, including not only the purely propositional ones, but also all those whose truth-set *does not shrink* as a consequence of the augmentation operation:

Proposition 6.2.1

$$\Vdash \varphi \rightarrow [*]\,\varphi \quad \text{implies} \quad \Vdash \mathrm{K}_{Ex}\,\varphi \rightarrow \mathrm{K}_{Im}\,\varphi$$

Proof See that of Velázquez-Quesada (2013), Proposition 2 plus awareness-of. □

The counterexample provides a hint on why the rest of the formulas fail: they express not ontic facts, but rather *epistemic* ones and, in particular *negative awareness that* situations. Indeed, $\neg\,\mathrm{A}^{\mathrm{t}}\,q$ expresses that the agent has not acknowledged q as true, and then $\mathrm{A}^{\mathrm{t}}\,\neg\,\mathrm{A}^{\mathrm{t}}\,q$ says that the agent has acknowledged this. In other words, and considering her awareness-of, the agent knows explicitly that she does not know q explicitly.

However, she might have enough information to 'extract' what she currently knows she does not have. In the provided model (with $A := \{p, q\}$), at w_1, she knows explicitly both p and $p \rightarrow q$ (she is aware of all atoms, hence of all formulas, and both $[\![p]\!]^M = \{w_1, w_3\}$ and $[\![p \rightarrow q]\!]^M = \{w_1, w_3, w_4\}$ are in $N(w_1)$); thus, after deductive reasoning, she will realise that q is indeed the case, hence knowing q explicitly. But then, she will automatically stop acknowledging (and thus stop knowing explicitly) that she did not know q explicitly. In other words, she might know explicitly that she does not know q, but such *high-order* knowledge will be gone once she gets to know that q is indeed the case, i.e. she can acknowledge her own ignorance, but after performing all possible inferences she will not be any more ignorant, thus, she will not be able to acknowledge it.

The reason for the failure of the 'explicit is implicit' property is that the agent has knowledge not only about propositional facts but also about her own (and eventually other agents') knowledge. This knowledge (semantically, the neighbourhood function) changes through the augmentation operation: the agent might know something explicitly at some point, and yet not know it explicitly afterwards (semantically, the awareness-that component is the key: I can have $U \in N(w)$ with $U = [\![\varphi]\!]^M$ for some φ, but even though this implies $U \in N^*(w)$, nothing guarantees $U = [\![\varphi]\!]^{M^*}$). This is nothing but an instance of the so called 'Moorean phenomena' in *DEL*, which occurs when an epistemic action 'invalidates' itself. In its best known incarnation, this phenomenon appears as formulas that become false after being truthfully announced (van Ditmarsch and Kooi 2006; Holliday and Icard III 2010); here, it appears as formulas that stop being known after a deductive inference.

6.2.3 *Formal Alternatives for Representing the Epistemic Concepts*

Though in Chaps. 2 and 3 I already reviewed the concepts of 'awareness' and 'knowledge', respectively, discussing different proposals and alternative views, I have not entered into a concrete formal comparison. Hence, this subsection is devoted to point out some formal alternatives for representing the main epistemic concepts of the EAK-Schema. Though, as explained above, the main concept is EAK, its components are awareness-of and awareness-that. I will focus on what these two types of awareness represent for performing the following comparison with some contemporary proposals, highlighting the relationship to the present formalisation.

Awareness-of. As is well known, this concept describes what the agent entertains, and it does not imply any attitude in favour or against the given formula. Here, the concept is syntactically represented by means of a set of atomic propositions (in the style of Fagin and Halpern 1988), indicating the atoms the agent is currently aware of.

There are other alternatives for representing such notion. One of them is the treatment of the concept of *topics* in Berto and Hawke (2018) (cf. Berto 2018), with the topic of a sentence being what the sentence is *about* (Yablo 2014; cf. Fine 2016). Such proposal assumes an underlying set of topics $\mathcal{T}$, a topic *fusion* operation $\oplus : \mathcal{T} \times \mathcal{T} \rightarrow \mathcal{T}$ (intended to make topics part of bigger topics), and a topic *parthood* relation $\leq \subseteq (\mathcal{T} \times \mathcal{T})$ (defining a 'subtopic' partial order). The fusion operator $\oplus$ is assumed to be $(\mathbf{1}_{\oplus})$ idempotent ($t \oplus t = t$ for all $t \in \mathcal{T}$), $(\mathbf{2}_{\oplus})$ commutative ($t \oplus s = s \oplus t$ for all $t, s \in \mathcal{T}$) and $(\mathbf{3}_{\oplus})$ associative ($(t \oplus s) \oplus r = t \oplus (s \oplus r)$ for all $t, s, r \in \mathcal{T}$); the parthood relation $\leq$ is assumed to be $(\mathbf{1}_{\leq})$ reflexive ($t \leq t$), $(\mathbf{2}_{\leq})$ antisymmetric ($t \leq s$ and $s \leq t$ imply $t = s$) and $(\mathbf{3}_{\leq})$ transitive ($t \leq s$ and $s \leq r$ imply $t \leq r$), with the set of *atomic* topics $\mathcal{T}_0$ containing exactly those topics lying at the bottom of the $\leq$-ordering. The link from topics to formulas is given by a function $T : \mathcal{L} \rightarrow \mathcal{T}$, which assigns an atomic topic to atomic propositions (i.e. $T(p) \in \mathcal{T}_0$) and then uses the fusion operator to assign topics to more complex formulas ($T(\varphi) := \bigoplus_{p \in \mathrm{at}(\varphi)} T(p)$).

The theory of topics can be seen as a generalisation of *awareness-of based on atoms*: take a topic to be a set of atoms ($\mathcal{T} := \wp(\mathrm{P})$), topic fusion as the set-theoretical union ($\oplus := \cup$) and the parthood ordering as the subset ordering ($\leq := \subseteq$). This satisfies the indicated properties: for any $Q, R, S \in \wp(\mathrm{P})$, $(\mathbf{1}_{\oplus})$ $Q \cup Q = Q$ (idempotency), $(\mathbf{2}_{\oplus})$ $Q \cup R = R \cup Q$ (commutativity) $(\mathbf{3}_{\oplus})$ $(Q \cup R) \cup S = Q \cup (R \cup S)$ (associativity); moreover, there also is $(\mathbf{1}_{\leq})$ $Q \subseteq Q$ (reflexivity), $(\mathbf{2}_{\leq})$ $Q \subseteq R$ and $R \subseteq Q$ imply $Q = R$ (antisymmetry) and $(\mathbf{3}_{\leq})$ $Q \subseteq R$ and $R \subseteq S$ imply $Q \subseteq S$ (transitivity). Thus, taking $T(\varphi) := \mathrm{at}(\varphi)$ fulfils the desiderata.

There are also more semantic options for representing awareness-of. One alternative within the possible worlds realm is to use an equivalence relation, the *issue relation*, which creates a partition of the domain, standing for the 'degree of abstraction' the agent has according to the issues that are currently under consideration (i.e. the possibilities she is currently entertaining). For example, consider a model with four possible worlds, exactly the four possible valuations for atoms in $\{p, q\}$. A 'blissfully ignorant' agent (one not knowing anything about these atoms, and yet being aware of both and also fully introspective about her own knowledge) sees the model as four different possibilities, i.e. four partitions, each one of them containing a single possible world. However, an agent only aware of p is not considering the further differences q can make. Thus, she sees the model as only two partitions: one in which p holds (hence containing the worlds $[p, q]$ and $[p, \neg q]$), and another in which p fails (hence containing the worlds $[\neg p, q]$ and $[\neg p, \neg q]$); something analogous happens when an agent is only aware of q. Finally, an agent unaware of both p and q sees a single partition containing all possible worlds.

The issues under consideration do not need to be atoms: an agent pondering whether $p \rightarrow q$ holds will see two partitions, one in which the implication is true (with worlds $[p, q]$, $[\neg p, q]$ and $[\neg p, \neg q]$) and another in which the implication is false (with world $[p, \neg q]$). This relational representation of the issues at stake, used in logical frameworks in, e.g., Grossi (2009), van Benthem and Minică (2012), Baltag et al. (2018), is already present in different places dealing with the notion of questions, ranging from linguistics (Groenendijk and Stokhof 1997) to learning theory (Kelly 1996).

For the present proposal I have chosen a syntactic representation (a set of atoms) because it allows a finer distinction. For example, there are models in which two different atoms happen to be true in exactly the same possible worlds; in such cases, an issue relation would not allow agents that are aware of the first but not of the second, something a set of atoms can.

Awareness-that. As already known, awareness-that is intended to describe what the agent has accepted (i.e. acknowledged) as true. In frameworks dealing with non-omniscient agents without using the notion of 'awareness-of' (e.g., Konolige 1984; Levesque 1984; Duc 1997; Artemov and Nogina 2005; Jago 2009; Velázquez-Quesada 2013), this is typically understood as explicit knowledge: what the agent actually has.

This notion is usually represented *syntactically* by a set of formulas (among the ones listed before: Konolige 1984; Duc 1997; Jago 2009). When no further constrain is imposed in such sets (as, e.g., the closure under logical consequence required by belief sets in the *AGM* belief revision of Alchourrón et al. (1985)), this representation avoids any form of logical omniscience as, e.g., no validity is required to be an element of the set, and having an implication and its antecedent does not guarantee the presence of the consequent. Still, such settings can be criticised for being too fine-grained as, e.g., the agent might know/believe $\varphi \vee \psi$ without knowing/believing $\psi \vee \varphi$.

Other options have a more *semantic* flavour. For example, Levesque (1984) relies on *situations*, semantic entities that might support the truth or falsity of a formula, but can also support neither and even both (cf. the truthmaker semantics of Fine 2017). A more 'traditional' approach is the logic of local reasoning of Fagin and Halpern (1988), which uses standard possible worlds but follows the idea that, rather than having a single system of beliefs, real agents have a number of distinct systems, with different ones working on different contexts (the *fragmentation/compartmentalization* of Davidson 1986; Egan 2008).

In the present proposal I relied on the use of neighbourhood semantics, thus using a purely semantic representation that still achieves the goal of making the agent non-ideal (see Sect. 6.2.1).[9] Being it a purely semantic representation, this proposal's awareness-that cannot distinguish between logically equivalent formulas. Still, this 'problem' is taken care of by the additional syntactic component the notion of Explicit Aware Knowledge requires: the awareness-of.

[9] In fact, the referred *logic of local reasoning* (Fagin and Halpern 1988) is equivalent to a neighbourhood semantics in which each neighbourhood is assumed to be non-empty and also closed under supersets, thus making the knowledge of the agent closed under conjunction elimination.

Before closing, it is worthwhile mentioning the existence of other options, which represent a notion intuitively similar to awareness-that, but do so by relying on further epistemic concepts. The syntactic notion of justification in *Justification Logics* has been analysed in Sect. 4.2, but there are also (purely semantic) alternatives in which the agent's knowledge or belief relies on concepts such as *evidence* (van Benthem and Pacuit 2011; Özgün 2017) and *arguments* (Shi et al. 2018b, 2017, 2018a). In such settings, the precise interplay between the different pieces of evidence (resp., the different arguments) and the way they define knowledge and/or beliefs determine the closure properties of the latter notions. Thus, under adequate relaxations, they might provide suitable frameworks for dealing with the Explicit Aware Knowledge of non-ideal agents (see, e.g., Balbiani et al. 2019; Bjorndahl and Özgün 2019).

6.3 Formalisation of Actions

The provided framework allows the representation of the epistemic state of non-ideal agents. Still, some authors Drapkin and Perlis (1986), Duc (1997), van Benthem and Velázquez-Quesada (2010), among others) have argued that solutions of this form are not completely adequate. First, an agent's (explicit) knowledge can be weakened in many ways, and there is no clear method to decide which restrictions are the appropriate ones. Second, such approaches do not look at the heart of the matter: they still describe the agent's epistemic state at a single stage, without looking at how such state is reached. What is needed is, then, not only a representation of non-ideal agents, but also a representation of the *actions* through which such an agent can change her epistemic state.

Recognising the role of epistemic actions is crucial. First, asking for additional requirements for a notion to be called 'explicit aware knowledge' might make the agent non-omniscient, but providing the epistemic actions that allow her to fulfil such requirements guarantees that she will not be defective or ignorant (in other words, she will be *rational*). Second, introducing the actions that lead to 'omniscient' states may demystify them. In Conan Doyle's detective stories, the explanation offered at the end turns Holmes' 'magical powers' into a sequence of observations and deductive acts, making the whole procedure *"elementary, my dear Watson"*.

This section completes and complements Sect. 5.2, where I presented five different epistemic actions, providing some intuitive examples, and defined its main theoretical features. In what follows I will repeat the diagrams presented in the mentioned section and provide a formalisation in the present model. I will also discuss some of the basic formal properties these actions entertain with respect to EAK.

6.3.1 Becoming Aware-Of

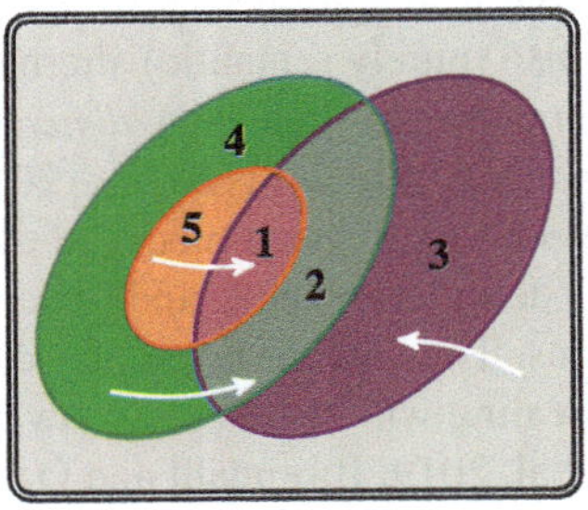

Actions of changes in awareness-of can be semantically represented by model operations that change the agent's atomic awareness set, leaving other model components unaffected (see, e.g., van Benthem and Velázquez-Quesada 2010; van Ditmarsch et al. 2012). The first one, becoming aware-of a given formula χ, boils down to adding χ's atoms to this set. Regarding the above diagram (from Fig. 5.7), the awareness-of ellipse (regions (1), (2) and (3)) will be extended with the atoms from χ.

Definition 6.3.1 (*Becoming aware-of operation*) Let $M = \langle W, N, V, A\rangle$ be an *ANM*; let χ be a formula in $\mathcal{L}$. The model $M^{+\chi} = \langle W, N, V, A^{+\chi}\rangle$ differs from M only in its atomic awareness set, which is extended with the atoms of χ:

$$A^{+\chi} := A \cup \text{at}(\chi)$$

For reasoning about the effects of such action, the language $\mathcal{L}_{[+\chi]}$ extends $\mathcal{L}$ with a *becoming aware-of* modality $[+\chi]$ for each formula χ: if φ is a formula in this extended language, then so is $[+\chi]\,\varphi$, a formula read as "*φ is the case after the agent becomes aware of χ*".[10] The modalities' semantic interpretation relies on the model the *becoming aware-of* operation yields[11]:

$$[\![[+\chi]\,\varphi]\!]^M := [\![\varphi]\!]^{M^{+\chi}}.$$

◀

Basic properties. The first property of this operation is simply a 'sanity check': after becoming aware of a formula χ, the agent is aware of it:

$$\Vdash [+\chi]\, \mathrm{A}^{\mathrm{o}}\, \chi$$

Moreover: awareness-of is based on atomic propositions, and thus can be seen as the language the agent has at her disposal. Then, becoming aware of a formula χ

[10]Define $\text{at}([+\chi]\,\varphi) := \text{at}(\chi) \cup \text{at}(\varphi)$.

[11]Defining a 'diamond' *becoming aware-of* modality in the standard way, $\langle+\chi\rangle\,\varphi := \neg\,[+\chi]\,\neg\varphi$, implies $[\![\langle+\chi\rangle\,\varphi]\!]^M = [\![\varphi]\!]^{M^{+\chi}}$: under this definition, $[+\chi]\,\varphi$ and $\langle+\chi\rangle\,\varphi$ are logically equivalent.

also makes the agent not only aware of all χ's atoms, but also aware of all formulas built from their combination. In particular, after becoming aware of a formula, the agent will be aware of all the formula's subformulas. E.g.,

$$\Vdash [+\neg\varphi]\,A^o\,\varphi, \qquad \Vdash [+A^o\,\varphi]\,A^o\,\varphi,$$
$$\Vdash [+(\varphi \land \psi)](A^o\,\varphi \land A^o\,\psi), \qquad \Vdash [+A^t\,\varphi]\,A^o\,\varphi,$$
$$\Vdash [+[*]\,\varphi]\,A^o\,\varphi.$$

More generally, after becoming aware of χ, the agent will be aware of φ if and only if she was already aware of all atoms that are in φ but not in χ[12]:

$$\Vdash [+\chi]\,A^o\,\varphi \leftrightarrow \bigwedge_{p \in \mathrm{at}(\varphi)\setminus\mathrm{at}(\chi)} A^o\,p$$

Its effect on Explicit Aware Knowledge. In other proposals for actions with similar spirit (see the mentioned references), becoming aware of an implicitly known formula gives the agent explicit knowledge about it. This is not the case here: the action makes the agent aware of the formula, but it does not guarantee that she will acknowledge it as true afterwards (the neighbourhood function is not affected by the operation). Even assuming that the agent has implicit knowledge of the formula before the *becoming aware-of* action is not enough, as this does not say anything about what the agent has acknowledged as true at such stage. Indeed, implicit knowledge only makes the agent aware of the formula (so the becoming aware-of action has no effect), but still the agent would need to do a further 'acknowledgement' step in order to make the formula part of her explicit knowledge.

- $\nVdash K_{Im}\,\chi \rightarrow [+\chi]\,A^t\,\chi$,
- $\nVdash K_{Im}\,\chi \rightarrow [+\chi]\,K_{Ex}\,\chi$.

Thus, becoming aware of a formula is not the same as recognising that it holds, i.e. being aware-that it is the case.

Given the stated $\Vdash [+\chi]\,A^o\,\chi$ and the fact that explicit knowledge is defined as awareness-of plus awareness-that, one might think that, if the agent has already acknowledged a given χ as true, then after becoming aware of it she would know it explicitly. This indeed is the case for propositional formulas

$$\Vdash A^t\,\gamma \rightarrow [+\gamma]\,K_{Ex}\,\gamma \qquad \text{for } \gamma \text{ a propositional formula}$$

and even for some non-propositional ones.[13] However, the formula is not valid in general, as the operation might not preserve acknowledgement of truth.

Fact 6.3.1 $\nVdash A^t\,\chi \rightarrow [+\chi]\,A^t\,\chi$.

---

[12] If there are no such atoms (if $\mathrm{at}(\varphi) \subseteq \mathrm{at}(\chi)$), the formula's right-hand side collapses to $\top$: after becoming aware of χ, the agent will be aware of every φ built only from atoms in χ.

[13] More precisely, the validity holds exactly for those formulas φ whose truth-set is not affected by the operation, that is, for every $\varphi \in \mathcal{L}$ for which $[\![\varphi]\!]^M = [\![\varphi]\!]^{M^{+\chi}}$.

Proof Take the set of atomic propositions $\{p\}$ and the formula $\varphi := \neg\,\mathrm{A^o}\,p$; consider the one-world model $M = \langle W = \{w\}, N, V, \varnothing\rangle$ with an empty atomic awareness set and a neighbourhood function given by $N(w) = \{W\}$ (the atomic valuation is not relevant). Note how $[\![\mathrm{A^o}\,p]\!]^M = \varnothing$ (since $p \notin A$), and therefore $[\![\neg\,\mathrm{A^o}\,p]\!]^M = W = \{w\}$; then,

$$[\![\mathrm{A^t}\neg\,\mathrm{A^o}\,p]\!]^M = \{u \in W \mid [\![\neg\,\mathrm{A^o}\,p]\!]^M \in N(u)\} = \{u \in W \mid W \in N(u)\} = \{w\}$$

so $w \in [\![\mathrm{A^t}\neg\,\mathrm{A^o}\,p]\!]^M$: at world w in model M, the agent has acknowledged that she is not aware of p. Now, the *becoming aware-of* operation with $\neg\,\mathrm{A^o}\,p$ yields an *ANM* $M^{+\neg\,\mathrm{A^o}\,p}$ whose atomic awareness set is $A^{+\neg\,\mathrm{A^o}\,p} = \{p\}$; thus, $[\![\mathrm{A^o}\,p]\!]^{M^{+\neg\,\mathrm{A^o}\,p}} = W$, and therefore $[\![\neg\,\mathrm{A^o}\,p]\!]^{M^{+\neg\,\mathrm{A^o}\,p}} = \varnothing$. Hence

$$[\![\mathrm{A^t}\neg\,\mathrm{A^o}\,p]\!]^{M^{+\neg\,\mathrm{A^o}\,p}} = \{u \in W \mid [\![\neg\,\mathrm{A^o}\,p]\!]^{M^{+\neg\,\mathrm{A^o}\,p}} \in N(u)\} = \{u \in W \mid \varnothing \in N(u)\} = \varnothing$$

so $[\![[+\neg\,\mathrm{A^o}\,p]\,\mathrm{A^t}\neg\,\mathrm{A^o}\,p]\!]^M = \varnothing$, that is, $w \notin [\![[+\neg\,\mathrm{A^o}\,p]\,\mathrm{A^t}\neg\,\mathrm{A^o}\,p]\!]^M$: at world w in the model that results from the agent becoming aware of $\neg\,\mathrm{A^o}\,p$, the agent has *not* acknowledged that she is not aware of p. Summarising,

$$w \notin [\![\mathrm{A^t}\neg\,\mathrm{A^o}\,p \rightarrow [+\neg\,\mathrm{A^o}\,p]\,\mathrm{A^t}\neg\,\mathrm{A^o}\,p]\!]^M.$$

□

The counterexample used to prove this fact shows again a Moorean effect: the agent might have accepted a formula as true, but a model operation might change the set of worlds in which the formula is true (in the counterexample, from W to $\varnothing$). Then, even though the neighbourhood function remains the same, the formulas the agent has acknowledged as true might change.

6.3.2 Becoming Unaware-Of

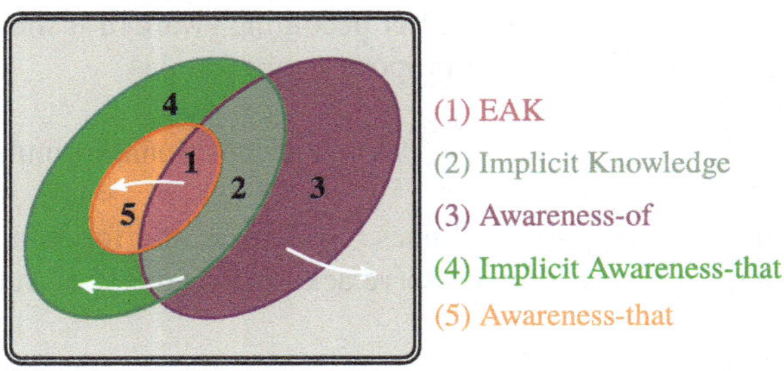

An agent can also become *unaware-of* a given formula; such action can be represented by shrinking her awareness set. In the above diagram (from Fig. 5.8) the arrows follow the exact opposite direction as in the previous action, i.e. the information moves outside the awareness-of (outside of regions (1), (2) or (3)).

Definition 6.3.2 (*Becoming unaware-of operation*) Let $M = \langle W, N, V, A\rangle$ be an *ANM*; let χ be a formula in $\mathcal{L}$. The model $M^{-\chi} = \langle W, N, V, A^{-\chi}\rangle$ differs from M only in its atomic awareness set, from which the atoms in χ have been removed:

$$A^{-\chi} := A \setminus \mathrm{at}(\chi)$$

On the syntactic side, $\mathcal{L}_{[-\chi]}$ extends $\mathcal{L}$ with a *becoming unaware-of* modality $[-\chi]$ for every formula χ. Formulas of the form $[-\chi]\varphi$, read as *"after the agent becomes unaware of χ, φ is the case"*,[14] are interpreted in the new model[15]:

$$[\![[-\chi]\varphi]\!]^{M} := [\![\varphi]\!]^{M^{-\chi}}.$$

◀

Basic properties. The operation passes the 'sanity check':

$$\Vdash [-\chi]\neg \mathrm{A}^{\mathrm{o}}\chi \qquad \text{for any } \chi \text{ such that } \mathrm{at}(\chi) \neq \varnothing.$$

The restriction is because the agent is always aware of every formula that does not involve atoms (e.g., $\top$, $\neg\top$, $\top \wedge \top$); thus, she cannot become unaware of any of them. Still, the agent becomes unaware of the given χ by becoming unaware of *all* its atoms (see below for a weaker alternative); thus, she also becomes unaware of *every formula involving atoms in χ*[16]:

$$\Vdash [-\chi]\neg \mathrm{A}^{\mathrm{o}}\varphi \qquad \text{for any } \varphi \text{ such that } \mathrm{at}(\chi) \cap \mathrm{at}(\varphi) \neq \varnothing.$$

Its effect on Explicit Aware Knowledge. Different from the *becoming aware-of* case, the *becoming unaware-of* operation behaves as expected in the sense that, after becoming unaware of a given formula (involving at least one atom), the agent will not have explicit knowledge about it, the reason being that, in such case, the agent will become unaware of the given formula, as observed before.

$$\Vdash [-\chi]\neg \mathrm{K}_{Ex}\,\chi \qquad \text{for any } \varphi \text{ such that } \mathrm{at}(\chi) \neq \varnothing,$$

***A weaker form of* becoming unaware-of**. The just defined action has very strong effects: after becoming unaware of χ, the agent also becomes unaware of every formula involving atoms in χ. Still, it is possible to provide an alternative operation, and then use it to define a weaker form of becoming unaware of.

Definition 6.3.3 (*Becoming unaware-of operation*) Let $M = \langle W, N, V, A\rangle$ be an *ANM*; take $\mathrm{Q} \subseteq \mathrm{P}$. The model $M^{-\mathrm{Q}} = \langle W, N, V, A^{-\mathrm{Q}}\rangle$ differs from M only in its atomic awareness set, from which the atoms in Q have been removed:

---

[14]Define $\mathrm{at}([-\chi]\varphi) := \mathrm{at}(\chi) \cup \mathrm{at}(\varphi)$.

[15]Just as before, defining $\langle -\chi\rangle\varphi := \neg[-\chi]\neg\varphi$ implies $\Vdash [-\chi]\varphi \leftrightarrow \langle -\chi\rangle\varphi$.

[16]In particular, by becoming unaware of χ, the agent becomes unaware of all χ's subformulas.

$$A^{-\mathrm{Q}} := A \setminus \mathrm{Q}$$

Syntactically, $\mathcal{L}_{[-\mathrm{Q}]}$ extends $\mathcal{L}$ with *becoming unaware-of* modalities $[-\mathrm{Q}]$ for $\mathrm{Q} \subseteq \mathrm{P}$, with $[-\mathrm{Q}]\,\varphi$ read as *"after the agent becomes unaware of atoms in* Q, φ *is the case"*[17] and semantically interpreted as[18]

$$[\![[-\mathrm{Q}]\,\varphi]\!]^{M} := [\![\varphi]\!]^{M^{-\mathrm{Q}}}.$$

Obviously, $\Vdash [-\chi]\,\varphi \leftrightarrow [-\mathrm{at}(\chi)]\,\varphi$. ◀

The operator $[-\mathrm{Q}]$ can be used to define another modality for an action of becoming unaware-of a formula by quantifying over its non-empty set of atoms:

$$[\sim\chi]\,\varphi := \bigwedge_{\{\mathrm{Q}\subseteq \mathrm{at}(\chi)\ \mid\ \mathrm{Q}\neq\varnothing\}} [-\mathrm{Q}]\,\varphi.$$

In words, φ is the case after the agent becomes unaware of χ if and only if it holds after the agent becomes unaware of any non-empty set of atoms of χ. This new modality also passes the 'sanity check', as

$$\Vdash [\sim\chi]\,\neg\,\mathrm{A}^{\mathrm{o}}\,\chi \qquad \text{for any } \chi \text{ such that } \mathrm{at}(\chi) \neq \varnothing.$$

Now, if χ involves atoms, the modality $[\sim\chi]$ is even stronger than $[-\chi]$. First, for such χs, the formula $[-\mathrm{at}(\chi)]\,\varphi$ occurs in the conjunction defining $[\sim\chi]$; thus,

$$\Vdash [\sim\chi]\,\varphi \rightarrow [-\chi]\,\varphi.$$

Second,

- $\not\Vdash [-\chi]\,\varphi \rightarrow [\sim\chi]\,\varphi$,

as φ holding after the agent becomes unaware of all atoms in χ does not guarantee that φ holds after she becomes unaware of any of its non-empty subsets.[19]

However, for similar χs (i.e. for χs involving atoms), the dual $\langle\sim\chi\rangle\,\varphi := \neg\,[\sim\chi]\,\neg\varphi$ is weaker. From its definition it follows that

$$\Vdash \langle\sim\chi\rangle\,\varphi \leftrightarrow \bigvee_{\{\mathrm{Q}\subseteq \mathrm{at}(\chi)\ \mid\ \mathrm{Q}\neq\varnothing\}} [-\mathrm{Q}]\,\varphi.$$

Then, first, since $[-\mathrm{at}(\chi)]\,\varphi$ occurs in the disjunction defining $\langle\sim\chi\rangle$,

$$\Vdash [-\chi]\,\varphi \rightarrow \langle\sim\chi\rangle\,\varphi.$$

---

[17]Define $\mathrm{at}([-\mathrm{Q}]\,\varphi) := \mathrm{Q} \cup \mathrm{at}(\varphi)$.

[18]Again, $\langle-\mathrm{Q}\rangle\,\varphi := \neg\,[-\mathrm{Q}]\,\neg\varphi$ implies $\Vdash [+\mathrm{Q}]\,\varphi \leftrightarrow \langle+\mathrm{Q}\rangle\,\varphi$.

[19]Take $\varphi := \bigwedge_{p\in\mathrm{at}(\chi)} \neg\,\mathrm{A}^{\mathrm{o}}\,p$, stating that the agent is unaware of *all* atoms in χ. Clearly, it holds when all atoms are removed, but fails when some of them remain.

Second, the fact that φ holds after removing some non-empty subset of atoms of χ from the agent's awareness set does not guarantee that it also holds after removing the specific subset $\mathrm{at}(\chi)$.[20] Thus,

- $\not\Vdash \langle \sim\chi \rangle \varphi \rightarrow [-\chi]\varphi$.

For a concrete case illustrating the relationship between these two modalities, take $\chi := p \wedge q$. The formula $[-(p \wedge q)](\mathrm{A}^{\circ} p \vee \mathrm{A}^{\circ} q)$ is not satisfiable, as the operation removes both atoms p and q. However, $\langle \sim(p \wedge q)\rangle(\mathrm{A}^{\circ} p \vee \mathrm{A}^{\circ} q)$ is, as there is a way to remove at least one atom from $p \wedge q$ that yields a model in which the agent is still aware of at least one of such atoms.

Despite describing a weaker form of becoming unaware of a formula, the modality $\langle \sim\chi \rangle$ works properly. Each one of the disjuncts that define it, $[-\mathrm{Q}]$ for Q a non-empty subset of $\mathrm{at}(\chi)$, guarantees that afterwards the agent will not be aware of χ ($\Vdash [-\mathrm{Q}] \neg \mathrm{A}^{\circ} \chi$ for every such Q, as the agent becomes unaware of at least one of χ's atoms); thus, after any of them, the agent will not know χ explicitly,

$$\Vdash \langle \sim\chi \rangle \neg \mathrm{K}_{Ex}\, \chi \qquad \text{for any } \varphi \text{ such that } \mathrm{at}(\chi) \neq \varnothing.$$

6.3.3 *Modus Ponens*

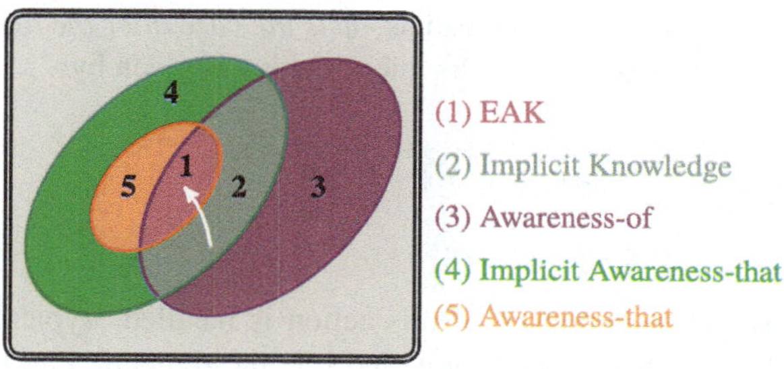

The two previous actions describe changes in what the agent entertains. Analogously, one can also represent actions that change what the agent has acknowledged as true, i.e. her awareness-that. The action of full deductive closure, represented by the augmentation operation, is already one of them; yet, it can be argued that such 'idealised' action is not best suited for the non-omniscient agent this proposal aims to describe.

A more appropriate action would be one that, instead of extracting at once all the consequences of what the agent currently has, extracts only a single piece of them (representing, thus, a deductive inference step). One could define different actions with such behaviour, as one performing a single act of conjunction introduction, or

[20]Take $\varphi := \bigvee_{p \in \mathrm{at}(\chi)} \mathrm{A}^{\circ} p$, stating that the agent is aware of at least one of χ's atoms: it holds when χ has at least two atoms and only one of them is removed, but it fails when all atoms are discarded.

one performing a single conjunction elimination step. I decided to stick to a more basic inference rule: *modus ponens*.[21]

The Modus ponens step, as well as every other deductive inference step, will move information from region (2) (implicit knowledge) to region (1) (Explicit Aware Knowledge) in the above diagram (from Fig. 5.9). This implies that the 'potential' information becomes explicit knowledge by the performance of the *Modus ponens* rule, taking for granted that the information is already part of the awareness-of of the agent. This action provides the deductive justification of knowledge in the EAK-Schema, as already discussed in Sects. 4.3 and 5.2.

Definition 6.3.4 (*Modus ponens operation*) Let $M = \langle W, N, V, A \rangle$ be an *ANM*; let $\xi \rightarrow \chi$ be an implication in $\mathcal{L}$. The model $M^{\xi \hookrightarrow \chi} = \langle W, N^{\xi \hookrightarrow \chi}, V, A \rangle$ differs from M only in the neighbourhood function, extending that of M with the truth-set of χ at M for those worlds where the agent knows explicitly both ξ and $\xi \rightarrow \chi$:

$$N^{\xi \hookrightarrow \chi}(w) := \begin{cases} N(w) \cup \{[\![\chi]\!]^M\} & \text{if } w \in [\![\mathrm{K}_{Ex}(\xi \rightarrow \chi) \wedge \mathrm{K}_{Ex}\,\xi]\!]^M \\ N(w) & \text{otherwise} \end{cases}$$

The operation adds $[\![\chi]\!]^M$ only to the neighbourhood of those worlds in which the agent knows explicitly both the implication $\xi \rightarrow \chi$ and its antecedent ξ. In this way, the operation truly represents an act of *modus ponens* inference.

For the syntax, $\mathcal{L}_{[\hookrightarrow]}$ extends $\mathcal{L}$ with a *modus ponens* modality $[^{\xi}\!\hookrightarrow_{\chi}]$ for each implication $\xi \rightarrow \chi$, with $[^{\xi}\!\hookrightarrow_{\chi}]\,\varphi$ read as "*φ is the case after the agent performs a modus ponens step with $\xi \rightarrow \chi$*".[22] Its interpretation is given by:

$$[\![[^{\xi}\!\hookrightarrow_{\chi}]\,\varphi]\!]^M = [\![\varphi]\!]^{M^{\xi \hookrightarrow \chi}}.$$

◀

Note the way the precondition of this action is handled. Typically, the operation always 'adds' the new information, and it is the semantic interpretation of its associated modality which takes care of applying it only when the information is truthful.[23] Here, it is the operation which takes care of adding the implication's consequent only when the agent knows explicitly both the implication and its antecedent, thus simplifying the semantic interpretation of $[^{\xi}\!\hookrightarrow_{\chi}]$.

Basic properties. Intuitively, after the truth-set of a given formula is added to the neighbourhood, the agent would acknowledge the formula as true, i.e. she would be aware that it is the case. However, Moorean phenomena also occurs in these

---

[21]Other proposals including operations with the same spirit are the deductive inferences of Velázquez-Quesada (2009), Velázquez-Quesada (2013), Solaki (2017), Smets and Solaki (2018), Balbiani et al. (2018), and the belief-based inference of Velázquez-Quesada (2014).

[22]Define $\mathrm{at}([^{\xi}\!\hookrightarrow_{\chi}]\,\varphi) := \mathrm{at}(\xi) \cup \mathrm{at}(\chi) \cup \mathrm{at}(\varphi)$.

[23]E.g., the case of a *public announcement* (Plaza 1989; Gerbrandy and Groeneveld 1997), whose precondition is for the announced formula to be true.

circumstances. Thus, after a formula's truth-set has been added to the neighbourhood, the agent might not acknowledge such formula as true, even if it is the consequent of an explicitly known implication whose antecedent is also explicitly known.

Fact 6.3.2 $\not\Vdash (\mathrm{K}_{Ex}(\xi \to \chi) \wedge \mathrm{K}_{Ex}\,\xi) \to [{}^{\xi}\!\hookrightarrow_{\chi}]\,\mathrm{A}^{\mathrm{t}}\,\chi$.

Proof Take the atoms $\{p, q\}$ and the *ANM* $M = \langle\{w_1, w_2, w_3\}, N, V, \{p, q\}\rangle$, with $V(p) = \{w_1\}$ $V(q) = \{w_1, w_2\}$, and

$$N(w_1) = \{\{w_1, w_2\}, \{w_1, w_3\}\}, \qquad N(w_2) = \{\{w_1\}\}, \qquad N(w_3) = \{\{w_1\}\}.$$

Take the implication $q \to (p \wedge \neg\,\mathrm{A}^{\mathrm{t}}\,p)$. From $\mathrm{at}(q \to (p \wedge \neg\,\mathrm{A}^{\mathrm{t}}\,p)) \subseteq \{p, q\}$ it follows that $[\![\mathrm{A}^{\mathrm{o}}(q \to (p \wedge \neg\,\mathrm{A}^{\mathrm{t}}\,p)) \wedge \mathrm{A}^{\mathrm{o}}\,q]\!]^M = \{w_1, w_2, w_3\}$; moreover, $[\![q]\!]^M = \{w_1, w_2\}$ and

$$\begin{aligned}[\![q \to (p \wedge \neg\,\mathrm{A}^{\mathrm{t}}\,p)]\!]^M &= [\![\neg q]\!]^M \cup [\![p \wedge \neg\,\mathrm{A}^{\mathrm{t}}\,p]\!]^M\\ &= [\![\neg q]\!]^M \cup ([\![p]\!]^M \cap [\![\neg\,\mathrm{A}^{\mathrm{t}}\,p]\!]^M)\\ &= \{w_3\} \cup (\{w_1\} \cap \{w_1\})\\ &= \{w_1, w_3\},\end{aligned}$$

imply together that $[\![\mathrm{A}^{\mathrm{t}}(q \to (p \wedge \neg\,\mathrm{A}^{\mathrm{t}}\,p)) \wedge \mathrm{A}^{\mathrm{t}}\,q]\!]^M = \{w_1\}$. Therefore, it follows that $[\![\mathrm{K}_{Ex}(q \to (p \wedge \neg\,\mathrm{A}^{\mathrm{t}}\,p)) \wedge \mathrm{K}_{Ex}\,q]\!]^M = \{w_1, w_3\}$.

Now, the model after a modus ponens with $q \to (p \wedge \neg\,\mathrm{A}^{\mathrm{t}}\,p)$ is such that

$$\begin{aligned}N^{q\hookrightarrow_{p\wedge\neg\mathrm{A}^{\mathrm{t}}p}}(w_1) &= \{\{w_1, w_2\}, \{w_1, w_3\}, \{w_1\}\},\\ N^{q\hookrightarrow_{p\wedge\neg\mathrm{A}^{\mathrm{t}}p}}(w_2) &= N^{q\hookrightarrow_{p\wedge\neg\mathrm{A}^{\mathrm{t}}p}}(w_3) = \{\{w_1\}\}.\end{aligned}$$

Thus, $[\![p \wedge \neg\,\mathrm{A}^{\mathrm{t}}\,p]\!]^{M^{q\hookrightarrow_{p\wedge\neg\mathrm{A}^{\mathrm{t}}p}}} = [\![p]\!]^{M^{q\hookrightarrow_{p\wedge\neg\mathrm{A}^{\mathrm{t}}p}}} \cap [\![\neg\,\mathrm{A}^{\mathrm{t}}\,p]\!]^{M^{q\hookrightarrow_{p\wedge\neg\mathrm{A}^{\mathrm{t}}p}}} = \{w_1\} \cap \varnothing = \varnothing$, so $[\![\mathrm{A}^{\mathrm{t}}(p \wedge \neg\,\mathrm{A}^{\mathrm{t}}\,p)]\!]^{M^{q\hookrightarrow_{p\wedge\neg\mathrm{A}^{\mathrm{t}}p}}} = [\![[{}^{q}\!\hookrightarrow_{p\wedge\neg\mathrm{A}^{\mathrm{t}}p}]\,\mathrm{A}^{\mathrm{t}}(p \wedge \neg\,\mathrm{A}^{\mathrm{t}}\,p)]\!]^M = \varnothing$.

Putting the two pieces together,

$$w_1 \notin [\![(\mathrm{K}_{Ex}(q \to (p \wedge \neg\,\mathrm{A}^{\mathrm{t}}\,p)) \wedge \mathrm{K}_{Ex}\,q) \to [{}^{q}\!\hookrightarrow_{p\wedge\neg\mathrm{A}^{\mathrm{t}}p}]\,\mathrm{A}^{\mathrm{t}}(p \wedge \neg\,\mathrm{A}^{\mathrm{t}}\,p)]\!]^M.$$

□

Still, the property holds for those implications whose consequent's truth-set is not affected by the operation (this includes all propositional formulas).

Proposition 6.3.1

$$\Vdash (\mathrm{K}_{Ex}(\xi \to \chi) \wedge \mathrm{K}_{Ex}\,\xi) \to [{}^{\xi}\!\hookrightarrow_{\chi}]\,\mathrm{A}^{\mathrm{t}}\,\chi \qquad \text{for } \chi \in \mathcal{L}_{[\hookrightarrow]} \text{ s.t. } \Vdash \chi \leftrightarrow [{}^{\xi}\!\hookrightarrow_{\chi}]\,\chi.$$

Proof Take $M = \langle W, N, V, A\rangle$ and $w \in W$ s.t. $w \in [\![(\mathrm{K}_{Ex}(\xi \to \chi) \wedge \mathrm{K}_{Ex}\,\xi]\!]^M$. Because of the latter, $M^{\xi\hookrightarrow_\chi}$ is such that $[\![\chi]\!]^M \in N^{\xi\hookrightarrow_\chi}(w)$. Then, by the required equivalence, $[\![[{}^{\xi}\!\hookrightarrow_{\chi}]\,\chi]\!]^M \in N^{\xi\hookrightarrow_\chi}(w)$, that is, $[\![\chi]\!]^{M^{\xi\hookrightarrow_\chi}} \in N^{\xi\hookrightarrow_\chi}(w)$. Thus, $w \in [\![\mathrm{A}^{\mathrm{t}}\,\chi]\!]^{M^{\xi\hookrightarrow_\chi}}$, i.e. $w \in [\![[{}^{\xi}\!\hookrightarrow_{\chi}]\,\mathrm{A}^{\mathrm{t}}\,\chi]\!]^M$, as required. □

Its effect on Explicit Aware Knowledge. From Fact 6.3.2, it is clear that

- $\not\Vdash (\mathrm{K}_{Ex}(\xi \rightarrow \chi) \wedge \mathrm{K}_{Ex}\,\xi) \rightarrow [{}^{\xi}\hookrightarrow_{\chi}]\,\mathrm{K}_{Ex}\,\chi$.

However, even though the modus ponens operation does not guarantee awareness-that, it does guarantee awareness-of,

$$\Vdash (\mathrm{K}_{Ex}(\xi \rightarrow \chi) \wedge \mathrm{K}_{Ex}\,\xi) \rightarrow [{}^{\xi}\hookrightarrow_{\chi}]\,\mathrm{A}^{\mathrm{o}}\,\chi,$$

the reason being that, by knowing the implication explicitly, the agent is aware of it, and thus also aware of the implication's consequent, a fact that is not affected by the operation (the awareness set does not change). This, together with Proposition 6.3.1, yields the following *dynamic* version of the famous *K* axiom:

$$\Vdash \mathrm{K}_{Ex}(\xi \rightarrow \chi) \rightarrow (\mathrm{K}_{Ex}\,\xi \rightarrow [{}^{\xi}\hookrightarrow_{\chi}]\,\mathrm{K}_{Ex}\,\chi) \quad \text{for } \chi \in \mathcal{L}_{[\hookrightarrow]} \text{ s.t. } \Vdash \chi \leftrightarrow [{}^{\xi}\hookrightarrow_{\chi}]\,\chi.$$

6.3.4 *Observation*

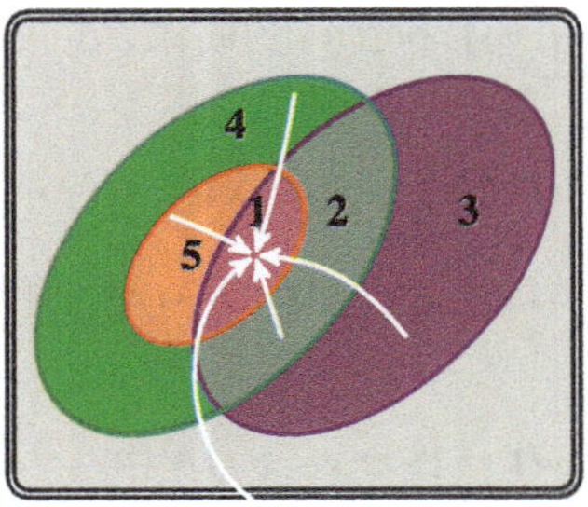

(1) EAK
(2) Implicit Knowledge
(3) Awareness-of
(4) Implicit Awareness-that
(5) Awareness-that

The actions studied so far, changes in awareness-of and two forms of deductive inference (the idealised full deductive closure described by [∗], and its just provided more realistic step-wise version, *Modus ponens*), are typically understood as 'internal' actions: what the agent entertains changes because her focus might change, and what she has accepted as true changes as a consequence of her own self-reflection. But there are also 'external' epistemic actions that reflect the way the agent interacts with her environment. A paradigmatic one is the act of receiving further information from an external source, what in *DEL* has been called a *public announcement*, as already discussed in several parts of this book. Since this model formalises a single-agent scenario, this action is considered here as an act of *observation*.

Intuitively, by observing (or receiving the information, from a trustful and truthful source) that a given χ is indeed the case, the agent becomes both aware of it (she entertains it) and aware that it is the case, meaning that she automatically acquires

Explicit Aware Knowledge of this formula. As the above diagram (from Fig. 5.11) depicts, the information can move from every region directly into region (1).

Definition 6.3.5 (*Observation operation*) Let $M = \langle W, N, V, A\rangle$ be an *ANM*; let χ be a formula in $\mathcal{L}$. The model $M^{!\chi} = \langle W, N^{!\chi}, V, A^{!\chi}\rangle$ differs from M on the neighbourhood function (extending that of M with χ's truth-set at M, when appropriate) and on the atomic awareness set (extending that of M with all atoms occurring in χ, as in the becoming aware-of operation):

$$N^{!\chi}(w) := \begin{cases} N(w) \cup \{[\![\chi]\!]^M\} & \text{if } w \in [\![\chi]\!]^M \\ N(w) & \text{otherwise} \end{cases}, \qquad A^{!\chi} := A \cup \text{at}(\chi).$$

Note that $[\![\chi]\!]^M$ is only added to the neighbourhood of those worlds in which χ holds. It is the action's precondition: for χ to be observed, it has to be the case. As specified in Sect. 5.2 the performance of this action constitutes the observational justification of knowledge in the EAK-Schema and it includes this precondition as a necessary requirement.

For the syntax, $\mathcal{L}_{[!\chi]}$ extends $\mathcal{L}$ with an *observation* modality $[!\chi]$ for each formula χ. Formulas of the form $[!\chi]\varphi$ are read as "*φ is the case after the agent observes χ*",[24] and are semantically interpreted as

$$[\![[!\chi]\varphi]\!]^M = [\![\varphi]\!]^{M^{!\chi}}.$$

Basic properties. The operation's 'sanity check' is a twofold task: one needs to check that, after observing χ, the agent not only becomes aware of χ, but also acknowledges that the formula is the case. The first part is straightforward, as the operation adds all atoms in χ to the agent's awareness set, and thus has, in this respect, the same effect as the becoming aware-of operation (Definition 6.3.1):

$$\Vdash [!\chi]\,\mathrm{A}^{\mathrm{o}}\,\chi$$

However, the second requirement does not hold:

Fact 6.3.3 $\nVdash [!\chi]\,\mathrm{A}^{\mathrm{t}}\,\chi$.

Proof As in Fact 6.3.1, take $\neg\,\mathrm{A}^{\mathrm{o}}\,p$ and the model $M = \langle W = \{w\}, N, V, \varnothing\rangle$ over the set of atoms $\{p\}$, with the neighbourhood function given by $N(w) = \{W\}$. As before, $[\![\neg\,\mathrm{A}^{\mathrm{o}}\,p]\!]^M = W$ (so $w \in [\![\neg\,\mathrm{A}^{\mathrm{o}}\,p]\!]^M$). But then, the model $M^{!\neg \mathrm{A}^{\mathrm{o}} p}$ that results from the observation of $\neg\,\mathrm{A}^{\mathrm{o}}\,p$ is such that $N^{!\neg \mathrm{A}^{\mathrm{o}} p} = \{W\}$ and $A^{!\neg \mathrm{A}^{\mathrm{o}} p} = \{p\}$. Then, $[\![\mathrm{A}^{\mathrm{o}}\,p]\!]^{M^{!\neg \mathrm{A}^{\mathrm{o}} p}} = W$ (from the second) and therefore $[\![\neg\,\mathrm{A}^{\mathrm{o}}\,p]\!]^{A^{!\neg \mathrm{A}^{\mathrm{o}} p}} = \varnothing$; hence,

$$\begin{aligned}[\![\mathrm{A}^{\mathrm{t}}\neg\,\mathrm{A}^{\mathrm{o}}\,p]\!]^{M^{!\neg \mathrm{A}^{\mathrm{o}} p}} &= \{u \in W \mid [\![\neg\,\mathrm{A}^{\mathrm{o}}\,p]\!]^{M^{!\neg \mathrm{A}^{\mathrm{o}} p}} \in N^{!\neg \mathrm{A}^{\mathrm{o}} p}(u)\} \\ &= \{u \in W \mid \varnothing \in N^{!\neg \mathrm{A}^{\mathrm{o}} p}(u)\} \\ &= \varnothing.\end{aligned}$$

[24] Define $\text{at}([!\chi]\varphi) := \text{at}(\chi) \cup \text{at}(\varphi)$.

Therefore, $[\![[!\neg \mathrm{A}^{\mathrm{o}} p]\, \mathrm{A}^{\mathrm{t}} \neg \mathrm{A}^{\mathrm{o}} p]\!]^{M} = \varnothing$: at world w in the model that results after the agent observes $\neg \mathrm{A}^{\mathrm{o}} p$, the agent has *not* acknowledged $\neg \mathrm{A}^{\mathrm{o}} p$. □

Just as before, this instance of Moorean phenomena should not be taken as a proof that the provided definition of the observation operation is incorrect. For those formulas χ whose truth-set is not affected by the operation (which, again, includes all purely propositional formulas), the expected property holds.

Proposition 6.3.2

$$\Vdash \chi \rightarrow [!\chi]\, \mathrm{A}^{\mathrm{t}} \chi \qquad \text{for } \chi \in \mathcal{L}_{[!\chi]} \text{ such that } \Vdash \chi \leftrightarrow [!\chi]\, \chi$$

Proof Let $M = \langle W, N, V, A\rangle$ be an *ANM*; take $w \in W$ with $w \in [\![\chi]\!]^{M}$. Because of the latter, $M^{!\chi}$ is such that $[\![\chi]\!]^{M} \in N^{!\chi}(w)$. Then, by the required equivalence, $[\![[!\chi]\, \chi]\!]^{M} \in N^{!\chi}(w)$, that is, $[\![\chi]\!]^{M^{!\chi}} \in N^{!\chi}(w)$. Thus, $w \in [\![\mathrm{A}^{\mathrm{t}} \chi]\!]^{M^{!\chi}}$, i.e. $w \in [\![[!\chi]\, \mathrm{A}^{\mathrm{t}} \chi]\!]^{M}$. □

Its effect on Explicit Aware Knowledge. From Fact 6.3.3, it follows that after the agent observes a formula χ she does not need to have explicit knowledge of it.

- $\nVdash [!\chi]\, \mathrm{K}_{Ex}\, \chi$.

However, from Proposition 6.3.2 and the way $[!\chi]$ affects what the agent entertains, the following validity holds:

$$\Vdash \chi \rightarrow [!\chi]\, \mathrm{K}_{Ex}\, \chi \qquad \text{for all } \chi \in \mathcal{L}_{[!\chi]} \text{ such that } \Vdash \chi \leftrightarrow [!\chi]\, \chi$$

6.3.5 *Forgetting*

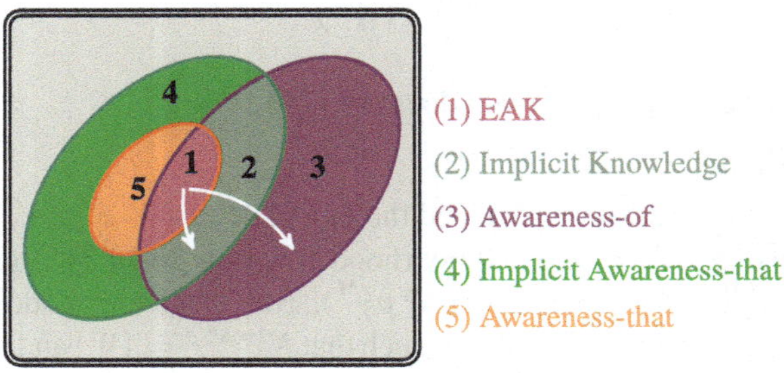

Finally, just as the agent can accept new information as true (via internal reasoning, performing a *Modus ponens* step, or external interaction, by observation), she can also 'forget' some of it, concretely, what she forgets is the truth of the previously acquired information.

This action is the awareness-that counterpart of the act of becoming unaware-of: what the agent entertains remains the same, but afterwards the agent might not be sure about the given formula's truth-value.[25] The above diagram's arrows (from Fig. 5.10) show how the Explicit Aware Knowledge (region (1)) will move to regions (2) or (3), remaining thus inside the awareness-of of the agent, but outside her awareness-that (outside of regions (1) and (5)).

Definition 6.3.6 (*Forgetting operation*) Let $M = \langle W, N, V, A\rangle$ be an *ANM*-model; let $\chi \in \mathcal{L}$ be a formula in $\mathcal{L}$. The model $M^{\backslash\chi} = \langle W, N^{\backslash\chi}, V, A\rangle$ differs from M only in its neighbourhood function, defined for $w \in W$ as

$$N^{\backslash\chi}(w) = N(w) \setminus \{[\![\chi]\!]^M\}$$

For the language, $\mathcal{L}_{[\backslash]}$ extends $\mathcal{L}$ with a modality for the *forgetting* operation: if χ and φ are formulas in the resulting extended language, then so is $[\backslash\chi]\,\varphi$, read as "*$\varphi$ holds after the agent forgets χ*",[26] and semantically interpreted as

$$[\![[\backslash\chi]\,\varphi]\!]^M = [\![\varphi]\!]^{M^{\backslash\chi}}$$

Basic properties. Intuitively, after the truth-set of a given formula is removed from the neighbourhood, the agent would not acknowledge the formula as true. However, just as in the previous cases, Moorean phenomena occurs: the agent might accept a formula as true, remove its truth-set from the neighbourhood, and yet still consider it true afterwards.

Fact 6.3.4 $\not\Vdash \mathrm{A^t}\,\chi \rightarrow [\backslash\chi]\,\neg\,\mathrm{A^t}\,\chi$.

Proof Take $\mathrm{P} = \{p\}$ and $M = \langle W = \{w_1, w_2\}, N, V, \varnothing\rangle$, with $V(p) = \{w_2\}$ and

$$N(w_1) = \{\{w_2\}, W\}, \qquad N(w_2) = \{\,\}.$$

Note how $[\![\neg\,\mathrm{A^t}\,p]\!]^M = W \setminus [\![\mathrm{A^t}\,p]\!]^M = W \setminus \{w_1\} = \{w_2\}$; hence, $[\![\neg\,\mathrm{A^t}\,p]\!]^M \in N(w_1)$ and thus $w_1 \in [\![\mathrm{A^t}\,\neg\,\mathrm{A^t}\,p]\!]^M$. However, $M^{\backslash\neg\mathrm{A^t}p}$ is such that

$$N^{\backslash\neg\mathrm{A^t}p}(w_1) = \{W\}, \qquad N^{\backslash\neg\mathrm{A^t}p}(w_2) = \{\,\}.$$

Thus, $[\![\neg\,\mathrm{A^t}\,p]\!]^{M^{\backslash\neg\mathrm{A^t}p}} = W \setminus [\![\mathrm{A^t}\,p]\!]^{M^{\backslash\neg\mathrm{A^t}p}} = W$; hence, $[\![\neg\,\mathrm{A^t}\,p]\!]^{M^{\backslash\neg\mathrm{A^t}p}} \in N^{\backslash\neg\mathrm{A^t}p}(w_1)$ so $w_1 \in [\![\mathrm{A^t}\,\neg\,\mathrm{A^t}\,p]\!]^{M^{\backslash\neg\mathrm{A^t}p}}$, i.e. $w_1 \in [\![[\backslash\neg\,\mathrm{A^t}\,p]\,\mathrm{A^t}\,\neg\,\mathrm{A^t}\,p]\!]^M$.

Therefore, $w_1 \in [\![\mathrm{A^t}\,\neg\,\mathrm{A^t}\,p \wedge [\backslash\neg\,\mathrm{A^t}\,p]\,\mathrm{A^t}\,\neg\,\mathrm{A^t}\,p]\!]^M$. □

As before, the expected property holds for those implications whose consequent's truth-set is not affected by the operation.

Proposition 6.3.3

$$\Vdash [\backslash\chi]\,\neg\,\mathrm{A^t}\,\chi \qquad \text{for } \chi \in \mathcal{L}_{[\backslash]} \text{ s.t. } \Vdash \chi \leftrightarrow [\backslash\chi]\,\chi$$

---

[25] See the relational-model-based *forgetting* operations of van Ditmarsch et al. (2009), Fernández-Duque et al. (2015).

[26] Define $\mathrm{at}([\backslash\chi]\,\varphi) := \mathrm{at}(\chi) \cup \mathrm{at}(\varphi)$.

Proof Take $M = \langle W, N, V, A \rangle$ and $w \in W$. From $N^{\backslash\chi}$'s definition, $[\![\chi]\!]^M \notin N^{\backslash\chi}(w)$; thus, from the equivalence, $[\![[\backslash\chi]\,\chi]\!]^M \notin N^{\backslash\chi}(w)$, that is, $[\![\chi]\!]^{M^{\backslash\chi}} \notin N^{\backslash\chi}(w)$, and therefore $w \notin [\![\mathrm{A^t}\,\chi]\!]^{M^{\backslash\chi}}$, so $w \in [\![\neg\,\mathrm{A^t}\,\chi]\!]^{M^{\backslash\chi}}$ and thus $w \in [\![[\backslash\chi]\neg\,\mathrm{A^t}\,\chi]\!]^M$. □

Its effect on Explicit Aware Knowledge. From Fact 6.3.4, it follows that after the agent forgets χ, she will not have Explicit Aware Knowledge of it,

- $\not\Vdash [\backslash\chi]\neg\,\mathrm{K}_{Ex}\,\chi$.

However, from Proposition 6.3.3,

$$\Vdash [\backslash\chi]\neg\,\mathrm{K}_{Ex}\,\chi \qquad \text{for } \chi \in \mathcal{L}_{[\backslash]} \text{ s.t. } \Vdash \chi \leftrightarrow [\backslash\chi]\,\chi$$

6.4 Conclusions on the Formal Model

This model proposes a formal setting defining the notion of *Explicit Aware Knowledge* based on two forms of awareness: awareness-of and awareness-that. The resulting notion of knowledge does not suffer from the idealisations that other notions of knowledge obtain under other semantic structures, as it separates the mere fact of entertaining some information (being aware of φ) from the acknowledgement that the information is indeed the case (being aware that φ holds). All these three notions are formally defined and their most important properties are discussed.

The proposed framework has other appealing features. First, it allows a natural definition of the notion of *implicit knowledge*, understood as the closure under logical consequence of its *explicit* counterpart, in terms of the well-known relationship between neighbourhood and relational models. Second, it also describes other epistemic notions that arise from combining the two mentioned forms of awareness, as what the agent has acknowledged as true but is not currently entertaining (informally, *disassociated* knowledge), or what she is not currently entertaining, and yet she can deduce from what she has acknowledged as true (*currently unreachable* knowledge). Third, it allows the representation of different epistemic actions, some of them affecting what the agent entertains, and some others affecting what she has accepted as true. These actions are formally defined and the way the agent's Explicit Aware Knowledge is affected by each one of them has been discussed.

There are several directions in which this model could be extended. On the technical side, the most important is a sound and complete axiom system axiomatising not only the basic framework (Sect. 6.1) but also its dynamic extensions (Sect. 6.3).[27]

On the conceptual side, an important next step is the move to the multi-agent setting, which will allow the representation of the information that 'real' agents have about the information of each other. Moreover, when several agents are involved, one can look at the public and private versions of the actions presented here, which will allow the agents to change their epistemic state in a way that might not be observed

[27] Still, the most important properties of the main concepts have been discussed.

by everyone. Finally, an appealing further direction is the extension to a setting in which the modelled agent has not only knowledge (in a more proper sense, probably satisfying the *truth axiom* $K\,\varphi \rightarrow \varphi$), but also beliefs. This would allow the discussion of the fine-grained versions of other epistemic actions, starting with *belief revision*, but also including forms of non-deductive inference (see Velázquez-Quesada 2014; Velázquez-Quesada 2015).

References

Alchourrón, C. E., Gärdenfors, P., & Makinson, D. (1985). On the logic of theory change: partial meet contraction and revision functions. *The Journal of Symbolic Logic, 50*(2), 510–530.

Artemov, S. N., & Nogina, E. (2005). Introducing justification into epistemic logic. *Journal of Logic and Computation, 15*(6), 1059–1073.

Balbiani, P., Fernández-Duque, D., Herzig, A., & Lorini, E. (2019). Stratified evidence logics. To appear in Proceedings IJCAI 2019.

Balbiani, P., Fernández-Duque, D., & Lorini, E. (2018). The dynamics of epistemic attitudes in resource-bounded agents. *Studia Logica*, Special Issue: 40 years of FDE, 1–32.

Baltag, A., Boddy, R., & Smets, S. (2018). Group knowledge in interrogative epistemology. In H. van Ditmarsch & G. Sandu (Eds.), *Jaakko Hintikka, Outstanding Contributions to Logic* (Vol. 12, pp. 131–164). Berlin: Springer.

Berto, F. (2018). Aboutness in imagination. *Philosophical Studies, 175*(8), 1871–1886.

Berto, F., & Hawke, P. (2018). Knowability relative to information. *Mind*, fzy045.

Bjorndahl, A., & Özgün, A. (2019). Uncertainty about evidence. *Electronic Proceedings in Theoretical Computer Science, 297*, 68–81.

Chellas, B. F. (1980). *Modal Logic: An Introduction*. Cambridge: Cambridge University Press.

Davidson, D. (1986). Deception and division. In J. Elster (Ed.), *The Multiple Self* (pp. 79–92). Cambridge: Cambridge University Press.

Drapkin, J. J. & Perlis, D. (1986). Step-logics: An alternative approach to limited reasoning. In *Proceedings of the European Conference on Artificial Intelligence* (pp. 160–163).

Duc, H. N. (1997). Reasoning about rational, but not logically omniscient agents. *Journal of Logic and Computation, 7*(5), 633–648.

Egan, A. (2008). Seeing and believing. perception, belief-formation and the divided mind. *Philosophical Studies, 140*(1), 47–63.

Fagin, R., & Halpern, J. Y. (1988). Belief, awareness, and limited reasoning. *Artificial Intelligence, 34*(1), 39–76.

Fernández-Duque, D., Nepomuceno-Fernández, Á., Sarrión-Morillo, E., Soler-Toscano, F., & Velázquez-Quesada, F. R. (2015). Forgetting complex propositions. *Logic Journal of the IGPL, 23*(6), 942–965.

Fernández-Fernández, C., & Velázquez-Quesada, F. R. (2019). A formal model for explicit knowledge as awareness-of plus awareness-that. In I. Sedlár & M. Blicha (Eds.), *The Logica Yearbook 2018* (pp. 101–115). London: College Publications.

Fernández-Fernández, C., & Velázquez-Quesada, F. R. (2020). Awareness of and awareness that: their combination and dynamics. *Logic Journal of the IGPL.*

Fine, K. (2016) Angellic content. *Journal of Philosophical Logic, 45*(2), 199–226.

Fine, K. (2017). Truthmaker semantics. In B. Hale, C. Wright, & A. Miller (Eds.), *A Companion to the Philosophy of Language* (Vol. 2, 2nd edn., pp. 556–577). Hoboken: Wiley Blackwell.

Gerbrandy, J., & Groeneveld, W. (1997). Reasoning about information change. *Journal of Logic, Language and Information, 6*(2), 147–169.

Groenendijk, J., & Stokhof, M. (1997). Questions. In J. van Benthem & A. G. T. Meulen (Eds.), *Handbook of Logic and Language* (pp. 1055–1124). Amsterdam: Elsevier Science Publishers.

Grossi, D. (2009). A note on brute versus institutional facts: Modal logic of equivalence up to a signature. In G. Boella, P. Noriega, G. Pigozzi, & H. Verhagen, (Eds.), *Normative Multi-Agent Systems*, number 09121 in Dagstuhl Seminar Proceedings. Dagstuhl: Schloss Dagstuhl—Leibniz-Zentrum fuer Informatik.

Grossi, D., & Velázquez-Quesada, F. R. (2015). Syntactic awareness in logical dynamics. *Synthese*, *192*(12), 4071–4105.

Holliday, W. H., & Icard, T. F., III. (2010). Moorean phenomena in epistemic logic. In L. D. Beklemishev, V. Goranko, & V. B. Shehtman (Eds.), *Advances in Modal Logic 8, papers from the eighth conference on "Advances in Modal Logic, held in Moscow, Russia, 24–27 August 2010* (pp. 178–199). London: College Publications.

Jago, M. (2009). Epistemic logic for rule-based agents. *Journal of Logic, Language and Information*, *18*(1), 131–158.

Kelly, K. T. (1996). *The Logic of Reliable Inquiry*. New York: Oxford University Press.

Konolige, K. (1984). *A Deduction Model of Belief and its Logics*. Ph.D. thesis, Computer Science Department, Stanford University.

Konolige, K. (1986). What awareness isn't: A sentential view of implicit and explicit belief. In *Proceedings of the 1st Conference on Theoretical Aspects of Reasoning about Knowledge, Monterey, CA, March 1986* (pp. 241–250).

Levesque, H. J. (1984). A logic of implicit and explicit belief. In R. J. Brachman, (Ed.), *Proceedings of the National Conference on Artificial Intelligence. Austin, TX, August 6–10 1984* (pp. 198–202).

Montague, R. (1970). Universal grammar. *Theoria*, *36*(3), 373–398.

Moss, L. S., de Queiroz, R. J. G. B., & Martínez, M., (Eds.). (2018). *Logic, Language, Information, and Computation—25th International Workshop, WoLLIC 2018, Bogota, Colombia, July 24–27 2018, Proceedings*, volume 10944 of *Lecture Notes in Computer Science*. Berlin: Springer.

Özgün, A. (2017). *Evidence in Epistemic Logic: A Topological Perspective*. Ph.D. thesis, Institute for Logic, Language and Computation (ILLC), Universiteit van Amsterdam, Amsterdam, The Netherlands. ILLC Dissertation series DS-2017-07.

Pacuit, E. (2017). *Neighborhood Semantics for Modal Logic*. Berlin: Springer.

Plaza, J. A. (1989). Logics of public communications. In M. L. Emrich, M. S. Pfeifer, M. Hadzikadic, & Z. W. Ras, (Eds.), *Proceedings of the 4th International Symposium on Methodologies for Intelligent Systems*, pp. 201–216. Tennessee: ORNL.

Scott, D. (1970). Advice on modal logic. In K. Lambert, (Ed.), *Philosophical Problems in Logic* (pp. 143–173). Dordrecht: D. Reidel.

Shi, C., Smets, S., & Velázquez-Quesada, F. R. (2017). Argument-based belief in topological structures. In J. Lang, (Ed.), *Proceedings Sixteenth Conference on Theoretical Aspects of Rationality and Knowledge, TARK 2017, Liverpool, UK, 24–26 July 2017.*, volume 251 of *EPTCS* (pp. 489–503).

Shi, C., Smets, S., & Velázquez-Quesada, F. R. (2018a). Beliefs based on evidence and argumentation. In Moss et al. (2018) (pp. 289–306).

Shi, C., Smets, S., & Velázquez-Quesada, F. R. (2018b). Beliefs supported by binary arguments. *Journal of Applied Non-Classical Logics*, *28*(2–3), 165–188.

Smets, S. & Solaki, A. (2018). The effort of reasoning: Modelling the inference steps of boundedly rational agents. In Moss et al. (2018) (pp. 307–324).

Solaki, A. (2017). Steps out of logical omniscience. Master's thesis, Institute for Logic, Language and Computation. ILLC Master of Logic Thesis Series MoL-2017-12.

van Benthem, J. (2011). *Logical Dynamics of Information and Interaction*. New York: Cambridge University Press.

van Benthem, J., & Minică, S. (2012). Toward a dynamic logic of questions. *Journal of Philosophical Logic*, *41*(4), 633–669.

van Benthem, J., & Pacuit, E. (2011). Dynamic logics of evidence-based beliefs. *Studia Logica*, *99*(1), 61–92.

van Benthem, J., & Velázquez-Quesada, F. R. (2010). The Dynamics of Awareness. *Synthese, 177*(Suppl.-1),5–27.
van Ditmarsch, H., French, T., & Velázquez-Quesada, F. R. (2012). Action models for knowledge and awareness. In W. van der Hoek, L. Padgham, V. Conitzer, & M. Winikoff (Eds.), *AAMAS 2012, Valencia, Spain, June 4–8, 2012* (pp. 1091–1098). Richland: IFAAMAS.
van Ditmarsch, H., Herzig, A., Lang, J., & Marquis, P. (2009). Introspective forgetting. *Synthese, 169*(2), 405–423.
van Ditmarsch, H., van der Hoek, W., & Kooi, B. (2008). *Dynamic Epistemic Logic*. Dordrecht: Springer.
van Ditmarsch, H. P., & Kooi, B. P. (2006). The secret of my success. *Synthese, 151*(2), 201–232.
Velázquez-Quesada, F. R. (2009). Inference and update. *Synthese, 169*(2), 283–300.
Velázquez-Quesada, F. R. (2013). Explicit and implicit knowledge in neighbourhood models. In D. Grossi, O. Roy, & H. Huang, (Eds.), *Logic, Rationality, and Interaction—4th International Workshop, LORI 2013, Hangzhou, China, October 9–12, 2013, Proceedings*, volume 8196 of *Lecture Notes in Computer Science* (pp. 239–252). Berlin: Springer.
Velázquez-Quesada, F. R. (2014). Dynamic epistemic logic for implicit and explicit beliefs. *Journal of Logic, Language and Information, 23*(2), 107–140.
Velázquez-Quesada, F. R. (2015). Reasoning processes as epistemic dynamics. *Axiomathes, 25*(1), 41–60.
Yablo, S. (2014). *Aboutness*. Carl G. Hempel Lecture Series. Princeton: Princeton University Press.

Chapter 7
Conclusions

Abstract This chapter draws some general conclusions on the different paths the present book has covered. It starts by reviewing the central concepts that constitute the first chapters, which have a descriptive flavour. Then, it resumes the importance of the EAK-Schema, alluding to the specific conclusions of it and recalls the formal model and its role as a practical outcome of the conceptual schema. Finally, a list of possible future developments of this research is provided.

Keywords Unity of Epistemology and Epistemic Logic · Epistemic awareness · Conceptual schema · Theoretical foundations · Logical correlate.

The main goal of this research has been to build a bridge between Epistemology and Epistemic Logic. In particular, between Awareness Justification Internalism (AJI) and the dynamic approaches to Awareness Logic. In the process of building this bridge I have analysed three core notions, namely, Awareness, Knowledge and Justification, which constitute the 'bricks' for building the bridge. The resulting EAK-Schema represents the bridge, that can be crossed in both directions: from Epistemology to Logic and from Logic to Epistemology.

Three core notions (that turn out to be five). The three concepts I have considered are *Awareness* in Chap. 2, *Knowledge* in Chap. 3 and *Justification* in Chap. 4. The selection of these concepts stems from an analysis of the chosen epistemological view, AJI. In AJI knowledge is defined as the justified true belief where the subject is aware of the grounds that justify her belief. A re-interpretation of this view, in terms more related to the logical theory that underlies the contemporary proposals of Dynamic Epistemic Logic, allows me to define 'knowledge' as the true justified information, where the agent (or subject) is aware of the justifying process. In the mentioned chapters I have first analysed each concept in Epistemology, secondly, reviewed its role in the different logical approaches and thirdly, advanced my interpretation in the EAK-Schema.

Regarding the notion of 'Awareness', I split it into two distinct concepts: *Awareness-of* and *Awareness-that*. The *Awareness-of*, on the one hand, represents the attention of the agent at the given moment of evaluation, i.e. considering one spe-

C. Fernández-Fernández, *Awareness in Logic and Epistemology*,
Logic, Epistemology, and the Unity of Science 52,
https://doi.org/10.1007/978-3-030-69606-1_7

cific epistemic scenario. This concept can be understood as the language the agent has at her disposal. I have argued that the logical interpretation of Awareness-of needs to be syntactic, meaning, a given list of formulas or propositions needs to be defined, since the 'focus' of the agent's attention may change depending on a variety of factors, that cannot be defined by any semantic function.

The *Awareness-that*, on the other hand, stands for the true information the agent has acknowledged. Being this so, this notion represents a type of knowledge that is not yet 'justified', but only one step away. In contrast to the previous type of awareness, I argued that in its logical interpretation the Awareness-that is a semantic concept, that is, it needs to be defined by a function that determines the truth of the formulas in a given model. This notion of awareness-that has been named 'explicit knowledge' in other proposals, but in my understanding there is still one missing piece, namely the combination with Awareness-of, in order to call this knowledge 'explicit'.

With respect to the concept of 'Knowledge' I established that the 'real' knowledge an epistemological view takes into account, and thus, the type of knowledge that my conceptual schema will take as its core, is *Explicit Aware Knowledge* (EAK). That the knowledge I am defining needs to be 'explicit', means that it represents the 'actual' knowledge a 'real agent' (or human subject) has at a given moment. The explicitness of the knowledge is an intrinsic feature to most logical proposals that overcome the problem of logical omniscience and want to model the knowledge of agents with limited reasoning abilities. While this is common I added an additional request, namely, that the 'agents' whose knowledge I am defining need to be human subjects understood from the point of view of AJI. There lies the reason for my explicit knowledge to also be 'aware'. The Awareness Requirement that AJI imposes on the subjects, that is, that they need to be aware of the grounds that justify their beliefs, made me take 'Awareness-of' as a requirement for my 'explicit knowledge'. It therefore follows that EAK means 'true information the agent has acknowledged and is considering at a given moment', that is, the combination of *Awareness-that and Awareness-of*.

While EAK represents the kernel of the EAK-Schema and the only 'real' knowledge regarding the epistemological interpretation, I have also defined an *Implicit Knowledge*, inspired by the different logical approaches. Implicit Knowledge is a logical construct, when analysed from the epistemological perspective, since there is no 'real' correlate of implicit (or potential) information. But when seen from the logical point of view, the concept is not only useful but also necessary. In the first approaches to Awareness Logic 'implicit knowledge' represented the standard knowledge from Epistemic Logic, which was closed under logical consequence and caused the agents to be logically omniscient. In the syntactic proposals the implicitness is obtained by a deductive closure of the explicit information. The main difference between these approaches lies in the concept they choose to be prior in their system, either the implicit or the explicit knowledge. I have argued that in the EAK-Schema the primary notion is EAK and the implicit knowledge is what can be deduced from it, in concrete, what can be reached by the performance of the action of 'deductive inference'. Though, as I said, there is no real correlate to this notion from the epistemological view, it is nevertheless necessary, since it provides a set of information that is reachable to the agent. As such, the implicit knowledge represents every information the agent can reach with deductive inferences.

The last core notion is that of *Justification*. I have argued that in the proposal of the EAK-Schema, the notion of justification is interpreted as a process given by the epistemic actions the agent performs for obtaining EAK. I have proposed two types of justifications, that can be classified depending on the specific epistemic action the agent carries out. If the agent obtains EAK through an act of 'deductive inference', that is, inferring her EAK by herself, then she will have a *deductive justification*. If, by contrast, she obtains her EAK from an external source, meaning that she has performed an action of 'observation', then her knowledge will have an *observational justification*.

While I have established two distinct justifications, the concept itself remains the same in both cases: the process through which the agent obtains her Explicit Aware Knowledge. I have argued that this notion of 'justification' stands in correlation to the Awareness Requirement of AJI. Since EAK is formed by both Awareness-that and Awareness-of, what justifies this knowledge is the epistemic action that the agent performs inside her aware information, meaning, that the 'grounds of what justifies her information' are represented by the action that turns her information into EAK. From the different actions that can be defined upon the EAK-Schema these two, deductive inference and observation, are the ones that can create 'new' EAK.

As advanced in the title of the previous paragraphs, though the core notions started out being three, they have turned out to be five and more fine-grained: Awareness-of, Awareness-that, Explicit Aware Knowledge (EAK), Implicit Knowledge and Justification. These notions constitute the 'bricks' for building the bridge from Epistemology to Epistemic Logic.

Crossing the bridge. I have already explained that the EAK-Schema constitutes the bridge between AJI and a dynamic interpretation of Awareness Logic (as the formal *ANM-model* I developed), but I have not insisted in the utility of the bridge for both research fields, that is, on what it brings with it to actually cross the bridge. Metaphorically, one could say that each time a notion crosses the bridge, it takes some of the bricks with it and thus widens the bridge on one of its sides. This means, that though the bridge is a 'small' theoretical framework that connects a total of five concepts with both their epistemological and logical interpretation, each time a new concept is considered and analysed 'through' this framework, the framework itself is augmented including now the new concept or new feature regarding one specific notion.

The possibility of the EAK-Schema for serving as a theoretical foundation for a new logical system is presented in Chap. 6 with the ANM-model (*Awareness Neighbourhood Model*). This formal model is based on the structure of the EAK-Schema and shows one of the possible logical outcomes this schema permits. Of course, there can be different semantic structures that capture the same notions. The formal model included here can serve as an example of the possible correlates.

The other direction of the bridge, namely from Epistemic Logic to Epistemology, has already been discussed, though not explicitly pointed out. Every re-interpretation of the epistemological concepts I have performed in order to construe the EAK-Schema represent the effects that the schema has on the epistemological theory. It constitutes, thus, a virtuous circle, since the reconsideration of the basic epistemological concepts is needed for the construction of the schema (or bridge) itself; but, once it is built, these same concepts serve for re-defining other epistemological theories that are related to AJI.

The new paradigm of the studies of language and information, which has as its central notion the 'information change', gives rise to the dynamic view on information (embodied in Dynamic Epistemic Logic). Within this paradigm, every knowable 'thing' is considered 'information' and the different attitudes of the agent towards this information (her informational attitudes) will determine the nature of this information, turning it into belief or knowledge. This re-consideration of the notions of 'belief' and 'knowledge' from Epistemology is one example of the beneficial effects of the bridge. The fact that the notion of 'belief' is re-interpreted in the epistemological definition leads to a new understanding, inspired by the dynamic view, where the agent has access to different forms of information and she determines, step by step, if this information is subject to become knowledge and why (providing a justification).

The EAK-Schema as a first and simple step. If one attends to all the possibilities I have chosen to not include in the EAK-Schema or to not mention and analyse in the previous chapters, the list might turn out to be at least as long as this book. Establishing a correlation between two wide-ranged research fields and, at the same time, keeping an overview of the selected concepts and theories that intervene is a complex task. My purpose while creating the EAK-Schema has been to keep every concept as simple as possible. Analogous to most logical proofs, where one searches for a 'minimal model' that proves the intended outcome, and afterwards expands it, I have proceeded using the less concepts that were needed for the construction of the intended bridge.

This being so, I state that the EAK-Schema does fulfil its main purpose of serving as a bridge between Epistemology and Epistemic Logic, though only between two specific areas, namely, Awareness Justification Internalism and a dynamic version of Awareness Logic, respectively. The conceptual schema provides a theoretical framework that permits a re-interpretation of the epistemological theory, and, at the same time, serves as a theoretical foundation for a new logic that takes the human subject from AJI as its prototypical epistemic agent. Such a logic is presented with the formal ANM-model.

It follows from the above that the creation of the EAK-Schema has re-connected Epistemology and Epistemic Logic as intended. The disconnection I reviewed in Sect. 1.3 has been inverted, at least, regarding these two specific areas.

Final remarks and further work. A proper conclusion not only has to resume the whole research and make a conclusive point, but should also point out the new lines of inquiry that can be followed as future work. My *conclusive point* can be summed up with the following statement: Epistemology and Epistemic Logic do, in fact, deal with the same core notions and with the EAK-Schema it is possible to establish a common theoretical framework that serves as a foundation for both areas.

With respect to the *further work* that can be carried out, there are several possible paths one may follow. Since the presented correlation connects two specific areas from Epistemology and Epistemic Logic, the natural next step would be to consider other related areas from both fields. Regarding Epistemology, it would be interesting to analyse some of the externalist proposals and unravel their core notions in comparison to the ones established here. There are also contemporary views that escape the traditional classification of internalism versus externalism and are worth considering.

Regarding the logical approaches there is a long list of proposals, developed in the past decades, that include interesting concepts, which are similar to the ones of Awareness-of and Awareness-that (as reviewed in Sect. 6.2.3). These proposals might be considered from the point of view of one specific epistemological view in order to create a different conceptual schema or extend the present one. In short, the methodology I have followed for the creation of the EAK-Schema can be extrapolated for connecting two different proposals in Epistemology and Epistemic Logic.

Index

C. Fernández-Fernández, *Awareness in Logic and Epistemology*, Logic, Epistemology, and the Unity of Science 52,
https://doi.org/10.1007/978-3-030-69606-1

Author Index

C. Fernández-Fernández, *Awareness in Logic and Epistemology*, Logic, Epistemology, and the Unity of Science 52,
https://doi.org/10.1007/978-3-030-69606-1

The manufacturer's authorised representative in the EU is Springer Nature Customer Service Centre GmbH, Europaplatz 3, 69115 Heidelberg, Germany. If you have any concerns regarding our products, please contact ProductSafety@springernature.com

Printed and bound by CPI Group (UK) Ltd, Croydon, CR0 4YY
30/09/2026
02225306-0008